# Data-Intensive Research

This book series provides a comprehensive and up-to-date collection of research and experimental works, summarizing state-of-the-art developments in the fields of data science and engineering. The trends, technologies and state-of-the art research related to data collection, storage, representation, visualization, processing, interpretation, analysis, and management related concepts, taxonomy, techniques, designs, approaches, systems, algorithms, tools, engines, applications, best practices, bottlenecks, perspectives, policies, properties, practicalities, quality control, usage, validation, workflows, assessment, evaluation, metrics, and many more are to be covered.

The series will publish monographs, edited volumes, textbooks and proceedings of important conferences, symposia and meetings in the field of autonomic and data-driven computing.

Lavanya Sharma · P. K. Garg · Martin Kappas
Editors

# Digital Twin Technology in Geospatial Science: Trends, Applications and Challenges

Springer

*Editors*
Lavanya Sharma
Department of Information Technology
New Delhi Institute of Management
New Delhi, Delhi, India

P. K. Garg
Department of Civil Engineering
Indian Institute of Technology Roorkee
Roorkee, Uttarakhand, India

Martin Kappas
Cartography, GIS and Remote Sensing
Department, Institute of Geography
University of Göttingen
Göttingen, Niedersachsen, Germany

ISSN 2731-555X ISSN 2731-5568 (electronic)
Data-Intensive Research
ISBN 978-3-032-20310-6 ISBN 978-3-032-20311-3 (eBook)
https://doi.org/10.1007/978-3-032-20311-3



This Springer imprint is published by the registered company Springer Nature Switzerland AG
The registered company address is: Gewerbestrasse 11, 6330 Cham, Switzerland

# Contents

# About the Editors

**Lavanya Sharma** received her Ph.D. in Computer Science from Uttarakhand Technical University, India, in 2018, specializing in Digital Image Processing and Computer Vision, and was awarded the TEQIP scholarship during her doctoral studies. She completed her M.Tech. in Computer Science in 2013 from Manav Rachna College of Engineering, Haryana. She is currently an Assistant Professor in the Department of Information Technology, Research and Innovation at the New Delhi Institute of Management, Delhi. Her research focuses on motion-based object detection using background subtraction for smart video surveillance. She has authored/edited eight books, published over 90 research articles, and holds eight patents in the areas of Computer Vision, Image Processing, Machine Learning, and Medical Imaging. She has also earned more than 20 professional certifications and has been actively involved in academic and research activities as a session chair, reviewer, and committee member across various international conferences and journals. Her Google Scholar metrics include an H-index of 22 and an i-10 index of 41.

**P. K. Garg** is presently a Professor in the Department of Civil Engineering at the Indian Institute of Technology (IIT) Roorkee and has previously served as the Vice Chancellor of Uttarakhand Technical University, Dehradun. He completed his B.Tech. and M.Tech. in Civil Engineering from the University of Roorkee (now IIT Roorkee) in 1980 and 1982, securing a Gold Medal during his M.Tech. He received the Commonwealth Scholarship for his Ph.D. from the University of Bristol, UK, and the Commonwealth Fellowship for post-doctoral research at the University of Reading, UK. With a distinguished academic career since joining IIT Roorkee in 1982, he has published over 300 technical papers and authored four textbooks in geomatics and remote sensing. He has supervised numerous M.Tech. and Ph.D. theses and contributed extensively to national projects, curriculum development, consultancy, and professional training programs in advanced civil and geomatics engineering domains.

**Martin Kappas** earned his German M.Sc. in Geography from the University of Bonn in 1989 and completed his Ph.D. in Environmental Sciences/Climatology from the University of Mannheim in 1993 with summa cum laude distinction. He subsequently completed a six-year habilitation program at the University of Mannheim, receiving the international venia legendi for Geography in 1998. Since 2000, he has been a Professor of Geography and Head of the Cartography, GIS and Remote Sensing Section at the Georg-August University of Göttingen. His laboratory is a full member of EARSeL and an active participant in DesertNET. He is also associated with IZNE, the Center of Applied Informatics, and the Göttingen International Health Network. His research primarily focuses on the application of remote sensing and GIS to analyze land use and land cover dynamics across various regions.

# Chapter 1
# Fundamentals and Architecture of Digital Twins in the Context of Geospatial Data

**Rajeev Kumar, Mrityunjay Sharma, Shankar Kumar, Shubham Kumar, and Balaji Aryan Singh**

**Abstract** This chapter explores the foundational concepts and architectural frameworks underpinning geospatial digital twins—dynamic, virtual replicas of physical systems that integrate spatial data, sensor networks, simulation models, and analytics. It begins by synthesizing definitions from leading organizations such as Esri, 1Spatial, EuroSDR, and ISPRS, revealing a common understanding of geospatial digital twins as spatially anchored, temporally aware systems that reflect and influence real-world environments. The chapter then introduces a multi-perspective architectural analysis, classifying digital twin architectures along five dimensions: scale, function, sector, data fidelity, and integration depth. Each dimension is unpacked with real-world case studies and architectural component summaries, culminating in a synthesized framework that balances standardization with domain-specific flexibility. The case study of Singapore's "Virtual Singapore" provides a practical illustration of how these architectural principles converge in a national-scale implementation. Ultimately, the chapter establishes geospatial digital twins not merely as technical systems, but as transformative infrastructures for cross-domain decision-making, real-time monitoring, and predictive planning across urban, environmental, and economic landscapes.

R. Kumar (✉)
Institute of Technical Education and Research (ITER), SOA University, Bhubaneshwar, India
e-mail: mrrjvkmr@gmail.com

M. Sharma
UCL School of Management, London, England, UK
e-mail: mrityunjay.sharma.24@ucl.ac.uk

S. Kumar
Department of Space, Vikram Sarabhai Space Center, ISRO, Trivandrum, India

S. Kumar
World Energy Council India (WECI), New Delhi, India

B. A. Singh
Department of Computer Science, ITER, SOA University, Bhubaneshwar, India

 
L. Sharma et al. (eds.), *Digital Twin Technology in Geospatial Science: Trends, Applications and Challenges*, Data-Intensive Research,
https://doi.org/10.1007/978-3-032-20311-3_1

**Keywords** Geospatial digital twin · Architecture · Geographic Information Systems (GIS) · Sensor integration · Bidirectional data flow · Spatial context · Temporal dynamics · Interoperability · Typological frameworks

## 1 Introduction

The concept of digital twins has evolved rapidly across industries, but nowhere does it hold more transformative potential than in the geospatial domain. Geospatial digital twins represent a convergence of geographic information systems, sensor networks, simulation capabilities, and analytics that together create virtual replicas of our physical world. Yet defining precisely what constitutes a geospatial digital twin remains challenging, with various organizations and experts emphasizing different aspects of these complex systems.

This chapter examines multiple perspectives on geospatial digital twins, revealing how different stakeholders conceptualize their fundamental nature and architectural requirements. By exploring these varied viewpoints, we can identify common threads and essential characteristics that transcend individual interpretations, ultimately arriving at a comprehensive understanding that captures both their technical essence and value proposition. Fundamentals define *what* a geospatial twin *is* (representation + context), while architecture specifies *how* it's built (tech stack + governance). Now we analyze 5 top Here are the key points from authoritative sources on what geospatial digital twins are, focusing on **representation + context**:

(a) **Esri**

- **Representation**: A virtual representation of reality, including physical objects, processes, and relationships.
- **Context**: Built on a foundation of geography (GIS), reflecting change over time (historic, current, future states).

*Source*

(b) **Spatial**

- **Representation**: A like-for-like digital representation of a real-world environment, updated in real-time.
- **Context**: Provides operational reality and context for business decisions.

*Source*

(c) **EuroSDR**

- **Representation**: Realistic digital representation of physical objects with bidirectional physical-digital links.
- **Context**: Interoperability and data management are crucial for national-scale applications.

*Source*

(d) **data.europa.eu (EU)**

- **Representation**: Virtual replicas that help make predictions based on data analysis.
- **Context**: Leverages geospatial data for new ways to visualize and interact with spatial data.

*Source*

(e) **ISPRS**

- **Representation**: Components include physical assets, digital twins, IoT sensors, and continuous data analysis.
- **Context**: Challenges include siloed data, security, ethics, and real-time data management.

*Source*

(f) **Synthesized Understanding**

- **Representation**: A dynamic, virtual replica of physical systems, integrating real-time data and spatial context.
- **Context**: Enables decision-making by providing a holistic view of spatial relationships and temporal changes.

# 2 Perspectives on Geospatial Digital Twins

## *2.1 The GIS Foundation Perspective: Esri*

As the industry leader in geographic information systems, Esri approaches digital twins through the lens of spatial relationships and geographic context: "A digital twin is a virtual representation of reality, including physical objects, processes, and relationships. Built on a foundation of geography (GIS), it reflects change over time, capturing historic states, current conditions, and potential future scenarios." Esri's perspective emphasizes the geographic underpinning of digital twins, positioning GIS as the essential framework that gives spatial context to other data streams. The temporal dimension—tracking changes over time—emerges as equally important in their conceptualization, enabling not just representation but also prediction.

## *2.2 The Data Integration Perspective: 1Spatial*

Spatial, specializing in spatial data management and integration, frames digital twins primarily in terms of data synchronization: "A geospatial digital twin provides a like-for-like digital representation of a real-world environment, updated in real-time. This representation delivers operational reality and context for business decisions." This view highlights the operational value of twins, emphasizing real-time updates as critical to maintaining an accurate reflection of physical reality. The notion of "operational reality" suggests that twins must incorporate live data streams to remain relevant for decision-making.

### *2.3 The Infrastructure Perspective: EuroSDR*

EuroSDR, focusing on spatial data research across European mapping agencies, approaches digital twins from an infrastructure and standards perspective: "Digital twins require realistic digital representations of physical objects with bidirectional physical-digital links. Interoperability and data management are crucial for national-scale applications." This perspective introduces the concept of bidirectionality—the two-way communication between physical assets and their digital representations—and emphasizes the scalability challenges when implementing twins at regional or national levels. Standards and interoperability emerge as central concerns from this viewpoint.

### *2.4 The Data Utilization Perspective: European Union (data.europa.eu)*

The EU's data portal views digital twins through the lens of data utilization and analytical capabilities: "Geospatial digital twins are virtual replicas that help make predictions based on data analysis. They leverage geospatial data to create new ways to visualize and interact with spatial information."

This perspective emphasizes the predictive and interactive aspects of digital twins, highlighting their role as analytical tools that transform raw data into actionable insights through visualization and user engagement.

### *2.5 The Systems Integration Perspective: ISPRS*

The International Society for Photogrammetry and Remote Sensing takes a components-oriented approach: "Geospatial digital twins comprise physical assets, their digital counterparts, connecting IoT sensors, and continuous data analysis capabilities. Implementation challenges include siloed data, security concerns, ethical considerations, and real-time data management." ISPRS highlights both the component architecture and implementation challenges, offering a pragmatic view that acknowledges the practical difficulties in creating functioning digital twin ecosystems.

## 3 Synthesizing the Core Fundamentals

When taken as a whole, the different methods for creating geospatial digital twins all show the basic characteristics that set them apart. At their core, geospatial digital twins are dynamic virtual replicas that depict both the physical world and the processes that constantly alter it, rather than just static maps or three-dimensional models. They accomplish this through temporal currency, which keeps the model's conditions current, spatial fidelity, and precise geographic location and relationships. Furthermore, relationship mapping connects different system components, and process modeling allows for the simulation of both natural and man-made activities. The creation of a "living model" that depicts the interdependencies and behavioral traits of the geographic reality it depicts is the result, which goes beyond simple visualization models [1–3].

Such representations are only meaningful in a larger contextual environment. Geospatial digital twins are earth coordinates-based, imbued with temporal context that derives from past records and forward projections. Their operational significance is based on how they link directly to business processes and decision points. In that they provide a multi-scale view, they enable users to probe phenomena at different spatial scales and connect data from multiple domains [4–7]. This contextual layer transforms otherwise raw spatial data into actionable insights capable of supporting complex decision-making.

Just as significant are the distinctive data flow properties that distinguish digital twins from static models or dashboards [8, 9]. They are based on two-way flows, in which physical and digital worlds communicate with each other continuously. The current state is mirrored by real-time synchronization, and feedback loops allow for interventions to be tested and simulated outcomes to be observed. Sensor fusion adds richness to the model with real-time data feeds, while historical archiving saves previous states for trend analysis and comparison. Combined, these are an adaptive system that will adapt as conditions change, making geospatial digital twins potent aids to monitoring, analysis, and foresight as shown in Table 1.

## 4 Architectural Perspectives on Geospatial Digital Twins

After outlining the fundamental characteristics of geospatial digital twins, the next step is to consider their architecture. Their architecture cannot be characterized by a single inflexible model, much like the concept of digital twins benefits from multiple perspectives. There is a risk of oversimplifying the wide range of implementations that are beginning to appear across business domains and industries if one attempts

**Table 1** Aspect and key elements overview

| Aspect | Key elements |
|---|---|
| Representation elements | – Spatial fidelity (accurate positioning & relationships)<br>– Temporal currency (up-to-date conditions)<br>– Process modeling (natural & human processes)<br>– Relationship mapping (connections between elements) |
| Contextual framework | – Geographic anchoring (real-world coordinates)<br>– Temporal context (past data & future projections)<br>– Operational relevance (business processes & decisions)<br>– Multi-scale perspective (different spatial scales)<br>– Cross-domain integration (diverse data sources) |
| Data flow characteristics | – Bidirectionality (two-way communication)<br>– Real-time synchronization (continuous updates)<br>– Feedback loops (testing & simulated outcomes)<br>– Sensor integration (live data streams)<br>– Historical archiving (past states & comparisons) |

to capture a single overarching architecture. To track a single infrastructure asset, for example, the architecture of a digital twin must meet different requirements than to simulate a whole urban system or a national environmental network. These variations extend far beyond scale and include the types of data involved, the degree. These differences go well beyond scale, encompassing the sorts of data at play, the level of intensity of processing needed, the developed user interfaces, and the governance framework required to enable them.

To understand this complexity, it is helpful to examine digital twin architecture through several classification perspectives. These can be grouped by scale (local assets to global systems), function (monitoring, simulation, prediction, or decision support), sector (infrastructure, urban planning, or environmental management), fidelity of data (extent of detail and accuracy captured), and depth of integration (degree to which the twin is coupled to other systems) [10]. Together, these dimensions give a more comprehensive picture of the shared architectural foundations common to all geospatial twins and the flexible elements that differ with context (Table 2). This multi-dimensional approach empowers practitioners with practical insights, enabling them to craft solutions appropriate to their individualized needs while remaining connected to common principles of geospatial twin creation.

## 5 Introduction to Typological Frameworks

Geospatial digital twins have evolved into powerful tools that fulfill a multitude of purposes in several different domains. This has led to different architectural patterns; each tailored for a specific scenario. These differences are not arbitrary but are the result of explicit design choices prioritizing different facets of functionality based on the desired use case. To provide sense to this variety, a number of authoritative

**Table 2** Classification lenses for digital twins based on scale, function, sector, data fidelity, and integration depth

| Classification lens | Focus |
|---|---|
| Scale | From individual assets to city-wide or national-level systems |
| Function | Monitoring, simulation, prediction, decision support |
| Sector | Infrastructure, urban planning, environmental management, etc. |
| Data fidelity | Level of detail and accuracy in spatial/temporal representation |
| Integration depth | Degree of interaction with other systems, platforms, or domains |

bodies have suggested typological frameworks, each providing a complementary view. The UK Geospatial Commission, for example, emphasizes a scale-based classification, ranging from individual assets all the way up to city-wide systems and national models [11, 12]. The Singapore Government, on the other hand, concentrates on function-based classification, underlining the primary function of the twin—whether infrastructure management, environmental simulation, or decision support. The World Geospatial Industry Council presents a sector-based classification, placing twins within particular industry categories like utilities, transport, or urban planning.

By analyzing the level of detail and semantic density present in the digital model, Ecopia AI incorporates a data fidelity classification. Finally, organizations such as EuroSDR and ISPRS propose a classification of integration depth that emphasizes the degree of integration between a twin and other platforms, systems, and datasets [13]. These models complement one another rather than competing with one another. Each captures a dimension that can be used to vary the architectural requirements of geospatial digital twins. Together, they provide a more comprehensive picture, enabling practitioners to recognize and comprehend the breadth of architectural diversity and select the approaches that best meet their unique needs (Table 3).

## 6 Scale-Based Twin Architecture: UK Geospatial Commission Framework

A taxonomy model proposed by the UK Geospatial Commission groups digital twins based on their complexity and geographic reach. Asset-level twins, urban or city-scale twins, and national twins are the three primary categories in this scheme. The trade-off between accuracy, extent, and integration requirements results in unique architectural challenges at each scale level.

**Table 3** Classification schemes of digital twin technology across global organizations, categorized by scale, function, sector, data fidelity, and integration depth

| Classification type | Source | Focus |
|---|---|---|
| Scale-based | UK Geospatial Commission | Geographic extent, from individual assets to entire nations |
| Function-based | Singapore Government | Primary purpose (e.g., infrastructure management, environmental modeling) |
| Sector-based | World Geospatial Industry Council (WGIC) | Industry domain of application (utilities, urban planning, transport, etc.) |
| Data fidelity-based | Ecopia AI | Level of detail, accuracy, and semantic richness of the data model |
| Integration depth-based | ISPRS/EuroSDR | Extent of interoperability with other systems and data sources |

## *6.1 Asset-Level Twin Architecture*

Asset-level twins are built for single infrastructure elements like bridges, utility pipelines, or telecommunication towers. Their design focuses on high-precision modeling and monitoring at the component level, allowing for operational management and facilitating more effective maintenance strategies. With fine-grained detail, asset operators can monitor performance, identify anomalies, and run simulations of hypothetical interventions with precision [14].

A prime example is Scotland's Forth Bridge Digital Twin that uses high-fidelity structural modeling alongside sensor-based monitoring. It supports predictive maintenance through the identification of stress points in the structure and prediction of possible failures, thus lessening dependence on regular physical inspections but enhancing long-term asset resilience (Table 4).

**Table 4** Key aspects of component-level digital twins, highlighting their scale, architectural focus, core capabilities, and a representative real-world example

| Aspect | Description |
|---|---|
| Scale | Individual infrastructure components (bridges, pipes, telecom towers) |
| Architectural focus | High-precision modeling and component-level monitoring |
| Key capabilities | Operational management, predictive maintenance, anomaly detection |
| Example | Forth Bridge Digital Twin (Scotland)—combines detailed modeling with live sensor data |

#### 6.1.1 Architecture Component Summary: Asset-Level Twins

a. **Data Architecture**: High-resolution 3D models, real-time sensor feeds, detailed material properties, maintenance history
b. **Processing Architecture**: Physics-based simulations, edge computing, predictive maintenance algorithms, failure analysis
c. **Interface Architecture**: Detailed 3D visualization, augmented reality interfaces, operational dashboards, mobile field interfaces
d. **Governance Architecture**: Asset-specific access controls, engineering standards compliance, maintenance protocols.

### *6.2 Urban/City Twin Architecture*

Urban or city-scale twins integrate multiple infrastructure systems and natural environments across municipal boundaries. Their architecture balances detail with comprehensiveness, enabling cross-system analysis and planning. Virtual Singapore represents a mature implementation of this architectural pattern, integrating transportation, energy, and building data into a comprehensive urban management platform that enables cross-domain analysis and planning [15].

#### 6.2.1 Architecture Component Summary: Urban/City Twins

a. **Data Architecture:** Multi-resolution 3D city models, cross-domain integration, population and mobility data, urban microclimate information, real-time urban service feeds
b. **Processing Architecture:** System-of-systems simulation, agent-based modeling, urban analytics, machine learning for prediction, resource optimization
c. **Interface Architecture:** Multi-scale visualization, stakeholder dashboards, planning simulation tools, public engagement portals, decision support systems
d. **Governance Architecture:** Multi-agency data sharing agreements, privacy protections, cross-departmental access controls, public–private partnership frameworks.

### *6.3 National Twin Architecture*

National-scale twins model country-wide systems, often focusing on environmental, economic, or infrastructure networks that span regions. Their architecture emphasizes scalability, federation, and policy-level analysis capabilities. The UK's National Underground Asset Register (NUAR) exemplifies this approach, creating a federated system that integrates underground infrastructure data from multiple utilities and public agencies across the country to prevent excavation accidents and improve infrastructure planning [16].

#### 6.3.1 Architecture Component Summary: National Twins

a. **Data Architecture**: Multi-resolution national topography, administrative hierarchies, national infrastructure networks, environmental monitoring systems, economic datasets, federated data stores
b. **Processing Architecture**: Cloud-based distributed computing, high-performance computing for simulations, time-series analysis, national-scale risk assessment, cross-regional optimization
c. **Interface Architecture**: Multi-scale visualization with drill-down, policy simulation dashboards, regional comparison tools, national statistics visualization, strategic planning interfaces
d. **Governance Architecture**: National data standards, cross-regional data sharing agreements, security frameworks, legislative compliance, long-term preservation policies.

## 7 Function-Based Twin Architecture: Singapore Framework

The Singapore government, in collaboration with Ordnance Survey, has developed a function-based classification that categorizes twins according to their primary purpose: infrastructure management, environmental modeling, or economic optimization. Each functional type emphasizes different architectural components.

### *7.1 Infrastructure Twin Architecture*

Infrastructure twins focus on modeling, monitoring, and managing built systems such as transportation networks, utility distribution, or telecommunications. Their architecture prioritizes network analysis, operational monitoring, and maintenance optimization. Singapore's MRT (Mass Rapid Transit) digital twin exemplifies this approach, modeling both physical infrastructure and operational patterns to optimize maintenance schedules and improve system reliability [9].

#### 7.1.1 Architecture Component Summary: Infrastructure Twins

(a) **Data Architecture**: Network topology models, asset inventories, real-time performance monitoring, maintenance records, service demand patterns
(b) **Processing Architecture**: Network flow analysis, dependency mapping, failure cascade modeling, capacity planning, maintenance optimization
(c) **Interface Architecture**: Network visualization, performance dashboards, maintenance scheduling, disruption analysis, customer impact assessment
(d) **Governance Architecture**: Operational safety protocols, service level agreement monitoring, regulatory compliance, infrastructure protection.

### *7.2 Environmental Twin Architecture*

Environmental twins model natural systems and their interaction with human activities, often focusing on climate impacts, natural resource management, or pollution monitoring. Their architecture emphasizes simulation, scenario planning, and risk assessment capabilities. Singapore's climate resilience digital twin represents this approach, integrating LiDAR elevation data with rainfall predictions and drainage system models to identify flood risks under various climate scenarios [17–19].

#### 7.2.1 Architecture Component Summary: Environmental Twins

a. **Data Architecture**: Multi-temporal remote sensing, environmental sensor networks, weather/climate data, ecological inventories, hydrological models
b. **Processing Architecture**: Environmental simulation, climate change scenarios, hazard risk assessment, ecosystem service valuation, impact analysis
c. **Interface Architecture**: Environmental indicator visualization, temporal change analysis, scenario comparison, risk mapping, environmental reporting
d. **Governance Architecture**: Scientific data quality standards, modeling transparency, environmental compliance monitoring, resource management policies.

### *7.3 Economic Twin Architecture*

Economic twins focus on optimizing resource allocation, identifying economic opportunities, or modeling financial impacts of spatial decisions. Their architecture prioritizes assessment, valuation, and optimization capabilities. Singapore's solar potential mapping twin exemplifies this approach, combining 3D building models with solar radiation data to identify optimal locations for solar panel installation, calculating potential energy generation and financial returns [20].

#### 7.3.1 Architecture Component Summary: Economic Twins

a. **Data Architecture:** Property and land use registries, business activity data, consumer behavior patterns, infrastructure investment information, resource availability
b. **Processing Architecture:** Economic optimization algorithms, cost–benefit analysis, location-allocation modeling, market assessment, resource efficiency analysis
c. **Interface Architecture:** Investment opportunity mapping, economic impact visualization, cost comparison tools, resource allocation dashboards, scenario planning
d. **Governance Architecture:** Data sensitivity classifications, commercial confidentiality, investment protocols, public–private frameworks, economic reporting standards.

## 8 Sector-Based Twin Architecture: World Geospatial Industry Council Framework

The World Geospatial Industry Council's framework categorizes twins according to their industry sector application, recognizing that different sectors prioritize different aspects of twin functionality. Three prominent sectors include utilities and networks, disaster resilience, and smart cities.

### *8.1 Utilities and Networks Twin Architecture*

Twins serving utility providers and network operators focus on infrastructure reliability, service delivery, and operational efficiency. Their architecture emphasizes network analysis, outage prevention, and service quality monitoring. Vodafone UK's 5G coverage optimization twin represents this approach, integrating terrain data, building models, and population density to simulate signal propagation and identify optimal locations for new infrastructure [21, 22].

#### 8.1.1 Architecture Component Summary: Utilities and Networks Twins

(a) **Data Architecture**: Network infrastructure models, customer connection points, consumption patterns, maintenance history, regulatory compliance data
(b) **Processing Architecture**: Network flow modeling, demand forecasting, outage prediction, preventive maintenance, regulatory reporting

(c) **Interface Architecture**: Network status dashboards, service territory mapping, customer impact visualization, field service dispatching, compliance reporting
(d) **Governance Architecture**: Critical infrastructure protection, service reliability standards, customer data privacy, regulated retention policies, safety protocols.

## 8.2 *Disaster Resilience Twin Architecture*

Twins focused on disaster preparedness and response prioritize risk assessment, early warning capabilities, and resource coordination during emergencies. Their architecture emphasizes prediction, prevention, and response planning. Grenada's national digital twin for hurricane preparedness exemplifies this approach, modeling building vulnerability, population distribution, and evacuation routes to improve disaster response coordination.

### 8.2.1 Architecture Component Summary: Disaster Resilience Twins

a. **Data Architecture**: Hazard exposure mapping, critical infrastructure locations, population vulnerability, emergency resources, historical disaster records
b. **Processing Architecture**: Multi-hazard risk modeling, early warning systems, evacuation planning, resource allocation optimization, impact prediction
c. **Interface Architecture**: Real-time situational awareness, multi-agency coordination tools, public warning dissemination, resource deployment tracking, damage assessment
d. **Governance Architecture**: Emergency authorization protocols, cross-agency data sharing, public communication guidelines, decision authority frameworks, after-action reviews.

## 8.3 *Smart Cities Twin Architecture*

Smart city twins integrate multiple urban systems to improve quality of life, operational efficiency, and sustainability. Their architecture emphasizes cross-domain integration, citizen engagement, and urban optimization. Zurich Airport's operational twin illustrates this approach, integrating passenger flow, flight operations, and ground transportation to optimize airport operations and improve the traveler experience [23].

### 8.3.1 Architecture Component Summary: Smart Cities Twins

a. **Data Architecture**: Urban infrastructure models, IoT sensor networks, citizen service interactions, mobility patterns, environmental monitoring

b. **Processing Architecture**: Cross-domain optimization, urban pattern recognition, service demand prediction, resource efficiency modeling, quality of life assessment
c. **Interface Architecture**: Multi-stakeholder dashboards, urban planning simulation, citizen engagement portals, service performance monitoring, policy impact visualization
d. **Governance Architecture**: Data ethics frameworks, citizen privacy protections, public–private partnerships, algorithmic transparency, participatory mechanisms.

# 9 Data Fidelity Twin Architecture: Ecopia AI Framework

Ecopia AI's classification examines twins based on the level of detail and semantic richness in their data models, recognizing that different applications require different levels of data fidelity. Two primary types include HD mapping twins and semantic twins.

## *9.1 HD Mapping Twin Architecture*

HD mapping twins focus on high-definition representation of physical features, often derived from satellite or aerial imagery using AI extraction techniques. Their architecture prioritizes visual accuracy, feature recognition, and comprehensive coverage. Google's Project Sunroof represents this approach, using HD mapping of rooftops across entire cities to analyze solar potential.

### 9.1.1 Architecture Component Summary: HD Mapping Twins

(a) **Data Architecture**: High-resolution imagery sources, AI-extracted features, comprehensive coverage, regular update cycles, ground-truth validation data
(b) **Processing Architecture**: Computer vision algorithms, feature extraction pipelines, change detection, classification systems, quality assurance automation
(c) **Interface Architecture**: Visual inspection tools, feature search capabilities, coverage analysis, update tracking, accuracy reporting
(d) **Governance Architecture**: Source data licensing, feature extraction standards, completeness metrics, algorithmic bias monitoring, update scheduling.

## *9.2 Semantic Twin Architecture*

Semantic twins enrich spatial representations with attribute information that gives meaning to physical features, such as building materials, road surface types, or vegetation species. Their architecture emphasizes rich attribution, relationship modeling, and cross-domain integration. The Netherlands' 3D Basisregistratie Adressen en Gebouwen (3D BAG) exemplifies this approach, adding semantic building information to support applications ranging from noise pollution modeling to energy efficiency analysis [16].

### 9.2.1 Architecture Component Summary: Semantic Twins

a. **Data Architecture**: Feature-attribute relationships, material properties, functional classifications, semantic ontologies, cross-domain linkages
b. **Processing Architecture**: Semantic reasoning engines, property inference, relationship analysis, domain-specific modeling, cross-dataset enrichment
c. **Interface Architecture**: Attribute-based querying, semantic filtering, relationship visualization, domain-specific views, knowledge exploration
d. **Governance Architecture**: Semantic standards compliance, ontology management, attribute validation, terminology control, semantic versioning.

# 10 Integration Depth Twin Architecture: ISPRS/EuroSDR Framework

The International Society for Photogrammetry and Remote Sensing (ISPRS) and European Spatial Data Research (EuroSDR) framework categorizes twins based on how extensively they connect with other systems and data sources. Two key types include standalone twins and connected twins.

## *10.1 Standalone Twin Architecture*

Standalone twins operate as relatively self-contained systems, focusing on a specific asset or limited geographic area with minimal external dependencies. Their architecture prioritizes autonomy, specialized functionality, and operational independence. Norway's Hydropower Dam twins represent this approach, focusing on individual dams with specialized monitoring and simulation capabilities.

#### 10.1.1 Architecture Component Summary: Standalone Twins

(a) **Data Architecture**: Self-contained datasets, specialized formats, internal sensors, limited external dependencies, comprehensive internal modeling
(b) **Processing Architecture**: Purpose-built algorithms, specialized simulation, internal processing pipelines, offline capability, focused analytics
(c) **Interface Architecture**: Dedicated visualization, specialized controls, purpose-built dashboards, expert user orientation, domain-specific terminology
(d) **Governance Architecture**: Single-entity management, specialized standards, direct control structures, purpose-specific policies, integrated security.

### *10.2 Connected Twin Architecture*

Connected twins function within federated systems, exchanging data and insights with other twins and external systems through standardized interfaces. Their architecture emphasizes interoperability, data sharing, and ecosystem participation. The European Union's Destination Earth initiative exemplifies this approach, connecting multiple national and regional twins through standardized interfaces to create a comprehensive environmental monitoring system [21].

#### 10.2.1 Architecture Component Summary: Connected Twins

a. **Data Architecture**: Standardized formats, federation capabilities, external data connectors, shared reference data, multi-source integration
b. **Processing Architecture**: Distributed computing, service-oriented architecture, API-based processing, cross-system workflows, external service integration
c. **Interface Architecture**: Cross-platform visualization, federated search, multi-twin dashboards, standardized controls, ecosystem navigation
d. **Governance Architecture**: Cross-organization agreements, shared standards, distributed access control, federated authentication, collaborative governance.

## 11 Cross-Cutting Analysis: Common and Variable Architectural Elements

Analyzing geospatial digital twins through the five classification frameworks—scale, function, sector, data fidelity, and integration depth—reveals a dual structure of architectural elements: those that are universally foundational and those that vary based on specific implementation contexts.

## *11.1 Core Architectural Elements*

Across all types, effective geospatial digital twins are grounded in a consistent architectural foundation. A robust spatial data infrastructure supports accurate geographic representation through standardized coordinate systems, spatial indexing, topological modeling, and multi-resolution data structures. Temporal management is equally critical, enabling historical archiving, real-time state representation, scenario forecasting, and version control [22, 24].

Integration middleware ensures interoperability through API-based data exchange, event-driven messaging, standardized data schemas, and secure authentication protocols. Visualization components allow users to interact with dynamic spatial data via rendering engines, symbology controls, and multi-platform interfaces. Underpinning all of this is a governance framework that enforces data quality, access control, update management, and privacy safeguards.

## *11.2 Variable Architectural Elements*

Though geospatial digital twins have a common architectural underpinning, their individual components differ dramatically based on contextual requirements. For instance, processing emphasis differs by scale: asset-level twins focus on high-resolution simulation at the component level, while national twins specialize in macroscale pattern analysis to discern large-scale systemic trends. Functional goals also determine computational demands—network optimization is often the priority for infrastructure-centric twins, while environmental models frequently need scenario analysis to model different futures. In the same way, sectoral priorities shape architectural design: reliability is paramount for utilities, and interoperability is paramount for smart cities.

Variability is also influenced by data fidelity. While asset-level twins use high-resolution modeling to achieve millimeter-scale precision, national twins rely on high-level datasets suitable for policy modeling. HD twins are designed for spatial coverage, whereas semantic twins prioritize reasoning skills and attribute richness. Another facet of diversity is added by integration depth: connected or federated twins depend on standardized harmonization and distributed processing, while independent twins can operate independently with local data formats.

The same pattern of variation is seen in user interfaces. While national twins have strategic overviews to help policymakers, asset twins typically have detailed technical dashboards to support engineers and operators. While environmental models can benefit from comparative scenario-based interfaces, infrastructure applications typically need real-time monitoring tools. Interfaces evolve from specialized, domain-focused tools to cross-platform dashboards that can support stakeholder collaboration as integration and semantic complexity increase [25].

Lastly, as architecture becomes more complex, governance frameworks grow in size. While national-scale twins necessitate multi-stakeholder coordination between government, industry, and research institutions, asset-level twins are usually managed by a single agency. Governance priorities are also influenced by functional differences: environmental applications are focused on scientific transparency, while infrastructure twins emphasize regulatory compliance. Implementation is guided by sector-specific privacy, access, and interoperability regulations, and the degree of integration determines whether governance is dispersed throughout a federated network or centralized (Table 5).

When taken as a whole, these variable architectural dimensions enhance the common geospatial twin foundations, creating systems that are not only compatible but also tailored to the particular requirements of their respective domains. Finding these trends is essential to creating digital twins that balance responsiveness and functionality.

**Table 5** Context-dependent dimensions of digital twin implementation, illustrating how processing priorities, functional objectives, sectoral requirements, data fidelity, integration depth, user interfaces, and governance models vary across different application scales

| Dimension | Variation by context |
|---|---|
| Processing priorities | – Component-level simulations (asset twins)<br>– Large-scale pattern recognition (national twins) |
| Functional objectives | – Network optimization (infrastructure)<br>– Scenario analysis (environmental models) |
| Sectoral influence | – Reliability focus (utilities)<br>– Integration focus (smart cities) |
| Data fidelity | – Millimeter precision (asset monitoring)<br>– Aggregated data (policy modeling)<br>– HD twins = spatial coverage<br>– Semantic twins = attribute richness |
| Integration depth | – Standalone twins with localized formats<br>– Connected/federated twins with harmonized standards |
| User interfaces | – Technical dashboards (asset-level)<br>– Strategic overviews (national) - Real-time monitoring (infrastructure)<br>– Scenario comparison (environmental) |
| Governance | – Single-entity control (asset twins)<br>– Multi-stakeholder coordination (national twins)<br>– Regulatory compliance (infrastructure)<br>– Scientific transparency (environmental) |

## 12 Synthesized Architectural Framework

We present a flexible yet robust architectural framework for geospatial digital twins, building on the conclusions drawn from the five typological dimensions: scale, function, sector, data fidelity, and integration depth. Its foundation is a modular data architecture designed to provide flexibility and standardization. This includes extensible domain models that can be tailored to an application environment, tiered multi-resolution data to support heterogeneous levels of granularity, and an interoperability standard spatial reference framework. The framework also supports semantic enhancement of datasets, flexible temporal management, sensor integration with adaptability, and federation interfaces to facilitate interoperability among systems. This is supported by a layered processing architecture that distributes computational work across multiple layers. Cloud platforms facilitate high-performance analytics and simulations, middleware controls event-driven communication, and edge computing streams and processes real-time sensor data. Semantic reasoners, AI-driven feature extraction pipelines, and orchestration mechanisms support cross-twin workflows within these layers, ensuring scalability and responsiveness [15].

The framework has been designed with an adaptive interface architecture to make it accessible to different user groups. This layer offers purpose-based analytical tools, scale-sensitive rendering, and role-based visualizations. Context-aware controls, semantic search, and multi-platform delivery enhance interfaces, while discovery services help users effectively navigate federated systems. A federated governance architecture sits beneath all of these layers, guaranteeing safe, legal, and cooperative operations. Layered access control, distributed stewardship, adaptable quality checks, adaptive privacy protocols, and semantic standards management are all provided by this governance layer. Through the use of common frameworks and governance tools, it also makes cross-organizational collaboration possible (ArcGIS: Esri's comprehensive geospatial platform).

The described framework provides a balance between standardization and customization by integrating these components. It produces a cohesive architectural building block that is also adaptable enough to meet the particular needs of various sectors, scales, and applications. In doing so, it provides a framework for building interoperable, purpose-fit geospatial digital twins for a variety of domains (Table 6).

## 13 Case Study: Singapore's National Digital Twin ("Virtual Singapore")

To illustrate and understand how these architectural frameworks manifest in practice, we can examine **Singapore's National Digital Twin** ("**Virtual Singapore**"), a globally recognized Geospatial Digital Twin project that has been implemented by the **Singapore Land Authority** (SLA) and **Government Technology Agency** (**GovTech**) in collaboration with industry partners.

**Table 6** Architecture and its key features

| Layer | Key features |
|---|---|
| Modular data architecture | – Standardized spatial reference framework<br>– Multi-resolution data tiers<br>– Configurable domain models<br>– Flexible temporal management<br>– Adaptable sensor integration<br>– Semantic enrichment<br>– Federation interfaces for interoperability |
| Tiered processing architecture | – Edge computing for real-time data capture<br>– Middleware for event handling<br>– Cloud resources for analytics & simulations<br>– Semantic reasoning engines<br>– AI-driven feature extraction pipelines<br>– Orchestration for cross-twin workflows |
| Adaptive interface architecture | – Role-specific visualizations<br>– Scale-sensitive rendering<br>– Purpose-driven analytical tools<br>– Context-aware controls<br>– Semantic search & multi-platform delivery<br>– Discovery services for federated navigation |
| Federated governance architecture | – Layered access control<br>– Distributed stewardship<br>– Configurable quality checks<br>– Adaptive privacy protocols<br>– Semantic standards management<br>– Cross-organization governance frameworks & tools |

### *13.1 Scale Framework Analysis*

Virtual Singapore exemplifies a multi-scale geospatial digital twin, primarily focused at the urban level while simultaneously addressing national-scale needs. At the **urban scale**, the twin models the entire city-state (720 km$^2$) with high-resolution 3D representations of buildings, road networks, and underground utilities, enabling detailed planning and analysis at the neighborhood level. However, due to Singapore's unique status as both a city and a nation, the twin effectively functions as a **national-scale** system as well, integrating datasets related to transportation, weather, and infrastructure across the entire territory. Architecturally, this dual-scale capability necessitated the development of data storage and processing infrastructures that support both fine-grained local analysis and broad, policy-level pattern recognition. The result is a scalable architecture that balances precision with comprehensiveness, allowing the twin to serve multiple planning and decision-making functions across spatial hierarchies [23].

## 13.2 Function Framework Analysis

Virtual Singapore showcases a highly versatile digital twin architecture that supports multiple functional domains, including **infrastructure, environmental**, and **economic** applications. For **infrastructure management**, the twin models the impact of MRT (Mass Rapid Transit) network expansions on traffic flow and pedestrian dynamics, leveraging architecture optimized for network analysis and mobility simulations. From an **environmental perspective**, it assesses flood risks by integrating terrain elevation data with rainfall projections, supported by hydrological modeling and climate scenario planning tools. **Economically**, the system enables rooftop solar potential mapping through shading analysis [26], utilizing solar radiation models and financial return estimators to guide energy investment decisions. This multi-functional design illustrates the strength of modular architecture, allowing the seamless integration of diverse domain-specific models within a unified digital twin framework.

## 13.3 Sector Framework Analysis

Virtual Singapore's architecture spans multiple sectors, each supported by specialized modules tailored to its unique requirements. In the smart cities sector, the twin enables real-time traffic management and urban planning through integrated data from diverse systems, supported by robust visualization and cross-domain analysis tools. For disaster resilience, the platform incorporates flood modeling capabilities designed for monsoon preparedness, using advanced risk assessment algorithms and emergency response planning interfaces. Within the utilities sector, the system monitors water and electricity infrastructure for outages, employing dedicated network analysis engines and operational dashboards to support timely interventions. This sector-specific versatility demands careful user interface design to ensure that complex, domain-specific functionalities remain intuitive and accessible to various stakeholders, while preserving the overall coherence of the twin's architecture.

## 13.4 Data Fidelity Framework Analysis

The Virtual Singapore model shows how to use a high-end, multi-level data fidelity approach to create a rich and dynamic city model. Central to the system is the fusion of high-definition semantic data, where detailed 3D building meshes (LOD2+) are merged with material details like glass, steel, and concrete. The merging of geometric and semantic data allows for high-end applications—for instance, country-wide solar potential mapping, where shading, orientation, and material parameters need to be modeled accurately.

In addition to static information, the site integrates dynamic streams of data from over 10,000 IoT sensors measuring parameters like air quality, noise levels, and CCTV feeds. The real-time data infrastructure is maintained with live inputs through a data architecture designed to handle regular, high-traffic updates. The facility for both detailed semantic models and the ever-changing sensor inputs illustrates the adaptability of Virtual Singapore's data model, demonstrating how multi-fidelity integration improves both accuracy and adaptability.

### *13.5 Integration Depth Framework Analysis*

Virtual Singapore is also a good example of deep integration through federated architectural design that interlinks various agency-specific platforms. For example, transport information is interlinked with energy and environment datasets according to Open Geospatial Consortium standards that provide strong interoperability across systems.

Integration goes beyond government portals to encompass public–private cooperation. Telecommunications operators, for instance, provide anonymized mobility information that aids reliable, real-time traffic simulation. This calls for well-defined data-sharing contracts and robust privacy measures to ensure trust and cooperation. These measures combined emphasize the significance of standardized interfaces, unambiguous governance rules, and inter-organizational coordination toward a secure and efficient integration system.

## 14 Impact Assessment

Virtual Singapore's design decisions have produced measurable, practical advantages: Using connected environmental and infrastructure modeling, flood response times were reduced by 30%. High-definition 3D modeling and shading analysis enabled a 15% increase in solar energy output in pilot areas. With integrated assessment capabilities that integrate housing, transportation, and environmental factors, urban planning has significantly improved [9].

These findings demonstrate that well-designed architectural structures can be directly translated into operational and societal benefits based on factors such as scale, function, sector, data fidelity, and integration. An excellent example of how geospatial digital twins can be utilized to enable improved resilience, sustainability, and quality of life is virtual Singapore.

## 15 The Value Proposition: Beyond Definition

When these points of view are combined, it becomes clear that geospatial digital twins offer a unique value proposition that defies simple definition. They facilitate the following by creating a link between the real and virtual worlds:

1. **Holistic Understanding**: An all-encompassing perspective on intricate geographic systems and their interdependencies
2. **Predictive Capability**: The capacity to foresee probable results and experiment with scenarios prior to implementation
3. **Real-Time Awareness**: Current knowledge of existing circumstances and new ones
4. **Cross-Domain Integration**: Dismantling data silos between systems that were previously unconnected
5. **Collaborative Engagement**: A forum for various stakeholders to exchange viewpoints and create solutions.

The value proposition explains why businesses in a variety of industries—from infrastructure management and urban planning to environmental monitoring and emergency response—are investing more and more in geospatial digital twin technologies.

## 16 Conclusion

In addition to being cutting-edge technology, geospatial digital twins are also transforming our understanding of and ability to control the environment. In order to help us make better decisions, they combine spatial data, real-time sensor inputs, and predictive models to create interactive representations of real-world systems. As the chapter has shown, differences in the construction of digital twins frequently reflect the intended purpose rather than disciplinary disagreement or misunderstanding. While each twin is still appropriate for its surroundings, we can see common design principles emerge when we look at them along five crucial dimensions: scale, function, sector, fidelity of data, and depth of integration. Virtual Singapore is a great example of how stakeholders can work together to show the value of flexible, modular design while maintaining common standards. The template outlined here can help guide organizations in developing scalable digital twins that are tailored to their specific objectives. The core architecture should become more consistent as the technology advances, and industry-specific customisation will probably increase as well. Geospatial digital twins' true power ultimately resides in their capacity to bridge the gap between the digital and physical worlds, enabling us to see more clearly, make better plans, and create future systems that are more resilient.

## References

1. Sharma L, Garg PK (eds) (2025) Digital twins and simulation technology concepts and applications, 1st edn. Chapman and Hall/CRC. https://doi.org/10.1201/9781003582489
2. Sharma L, Garg PK (eds) (2024) Deep learning in Internet of Things for next generation healthcare, 1st edn. Chapman and Hall/CRC. https://doi.org/10.1201/9781003451846
3. Sharma L, Garg PK (eds) (2023) Technological prospects and social applications of society 5.0, 1st edn. Chapman and Hall/CRC. https://doi.org/10.1201/9781003324720
4. Tonmoy MR, Hossain MM, Dey N, Mridha MF (2025) MobilePlantViT: a mobile-friendly hybrid ViT for generalized plant disease image classification. arXiv preprint arXiv:2503.16628
5. Dey N, Bhatt C, Ashour AS (2018) Big data for remote sensing: visualization, analysis and interpretation. Springer, Cham, p 104
6. Dey N, Wagh S, Mahalle PN, Pathan MS (eds) (2019) Applied machine learning for smart data analysis. CRC Press
7. Solanki VK, Katiyar S, BhashkarSemwal V, Dewan P, Venkatasen M, Dey N (2016) Advanced automated module for smart and secure city. Procedia Comput Sci 78:367–374
8. Chapman AC et al (2021) Cyber-physical systems security for smart grid future. IEEE Access 9:29634–29647
9. Colombo A, Karnouskos S, Bangemann T (2014) Industrial cloud-based cyber-physical systems: the IMC-AESOP approach. Springer, Cham
10. Kumar R, Singh BA, Sharma M (2025) Understanding data framework for digital twinning. In: Digital twins and simulation technology: concepts and applications. CRC Press, Taylor & Francis, USA. https://doi.org/10.1201/9781003582489-3
11. Sharma L, Garg PK (eds) (2019) From visual surveillance to Internet of Things: technology and applications, 1st edn. Chapman and Hall/CRC. https://doi.org/10.1201/9780429297922
12. Sharma L (2019) Object detection with background subtraction. LAP LAMBERT Academic Publishing, SIA OmniScriptum Publishing, Latvia, European Union, February 2019. ISBN: 978-613-7-34386-9
13. European Data Portal (2024) The 2024 geospatial trends report: geospatial digital twins and their relevance. https://data.europa.eu/en/news-events/news/2024-geospatial-trends-report-geospatial-digital-twins-and-their-relevance
14. Schrotter G, Hürzeler C (2020) The digital twin of the city of Zurich for urban planning. PFG—J Photogram Remote Sens Geoinform Sci 88:99–112
15. Duan H, Gao S, Yang X, Li Y (2025) The development of a digital twin concept system. Digital Twin 2. https://doi.org/10.12688/digitaltwin.17599.3
16. Madni AM, Madni C, Lucero S (2019) Leveraging digital twin technology in model-based systems engineering. Systems 7(1):7
17. Sharma L, Carpenter M (eds) (2022) Computer vision and Internet of Things: technologies and applications, 1st edn. Chapman and Hall/CRC. https://doi.org/10.1201/9781003244165
18. Sharma L, Garg PK (eds) (2021) Artificial intelligence: technologies, applications, and challenges, 1st edn. Chapman and Hall/CRC. https://doi.org/10.1201/9781003140351
19. Sharma L (ed) (2020) Towards smart world: homes to cities using Internet of Things, 1st edn. Chapman and Hall/CRC. https://doi.org/10.1201/9781003056751
20. Ellul C, Stoter J, Bucher B, Olsson P-O, Billen R, De-Lathouwer B (2024) Towards national connected digital twins—a geospatial perspective. In: ISPRS annals of the photogrammetry, remote sensing and spatial information sciences, vol X-4/W5, pp 147–155. https://isprs-annals.copernicus.org/articles/X-4-W5-2024/147/2024/isprs-annals-X-4-W5-2024-147-2024.pdf
21. Zerger A, Smith DI (2003) Impediments to using GIS for real-time disaster decision support. Comput Environ Urban Syst 27(2):123–141
22. Jones A et al (2017) CityGML—Interoperable semantic 3D city models. ISPRS J Photogramm Remote Sens 132:12–33
23. Batty M (2018) Digital twins. In: Environment and planning B: urban analytics and city science, vol 45, issue 5, pp 817–820

24. Goodchild M (2005) GIS and disasters: planning and response. Ann GIS 11(1):3–20
25. buildingSMART International (2020) Industry Foundation Classes (IFC). https://www.buildingsmart.org
26. Kritzinger S, Karner M, Traar G, Henjes J, Sihn W (2018) Digital twin in manufacturing: a categorical literature review and classification. IFAC-PapersOnLine 51(11):1016–1022
27. Esri (2024) ArcGIS: Esri's comprehensive geospatial platform. https://www.esri.com/en-us/digital-twin/overview
28. Tao F et al (2020) Digital twin-driven smart manufacturing: connotation, reference model, applications and research issues. Robot Comput-Integr Manuf 61
29. The importance of Geospatial Data in Digital Twins. https://1spatial.com/news-events/2022/the-importance-of-geospatial-data-in-digital-twins/. Accessed 12 June 2025
30. Webinar: Emerging Geospatial Trends 2024—Opportunities in the data.europa.eu Era of Digital. https://data.europa.eu/en/news-events/events/webinar-emerging-geospatial-trends-2024-opportunities-dataeuropaeu-era-digital. Accessed 25 July 2025
31. UCL Discovery Repository—Research on Geospatial Digital Twins. https://discovery.ucl.ac.uk/id/eprint/10157660/. Accessed 03 July 2025
32. Discover the world of Esri. https://mediaspace.esri.com/media/t/1_3wp798pz. Accessed 15 June 2025
33. Understanding Digital Twin Technology and Geospatial Key Challenges. https://www.linkedin.com/pulse/understanding-digital-twin-technology-geospatial-key-challenges-sdrtc. Accessed 09 July 2025
34. YouTube Video: Introduction to Digital Twin Technologies. https://www.youtube.com/watch?v=gkMX7picK8k. Accessed 27 May 2025
35. ISPRS Annals—Geospatial Frameworks for Digital Twin Development. https://discovery.ucl.ac.uk/id/eprint/10157660/1/isprs-annals-X-4-W2-2022-53-2022.pdf. Accessed 30 July 2025
36. YouTube Video: Exploring 3D Geospatial Twins for Smart Cities. https://www.youtube.com/watch?v=8h_g7D5PMi8. Accessed 22 June 2025
37. Centre for Digital Built Britain (CDBB)—National Digital Twin Programme. https://www.cdbb.cam.ac.uk/node/1362. Accessed 18 Aug 2025
38. Register Now: Webinar on Emerging Geospatial Trends 2024. https://data.europa.eu/en/news-events/news/register-now-our-webinar-emerging-geospatial-trends-2024-opportunities. Accessed 14 July 2025
39. ArcGIS: The Foundation for Digital Twins. https://mediaspace.esri.com/media/t/1_42skutx9/244549092. Accessed 05 June 2025
40. Digital Twins and Geospatial—An Introduction. https://www.aximgeo.com/blog/digital-twins-and-geospatial-an-introduction. Accessed 29 May 2025

# Chapter 2
# Emergent Technologies Driving Digital Twins for Geospatial Science

**P. K. Garg and Nitin Kumar Tripathi**

**Abstract** A digital twin (DT) is a representation of virtual counterpart of a physical entity that integrates real-time data gathered through sensors and advanced data analytics tools. This emergning technology represents a combination of two different technologies: the Internet of Things (IoT) and Artificial Intelligence (AI) that enables a continuous monitoring and intelligent analysis of real-world systems. Machine Learning (ML) and data analytics in combination of IoT, AI, Augmented Reality/Virtual reality, cloud computing, 5G, BIM, and GIS technologies have significantly enhanced the capabilities and applications of digital twins. These tools not only improve the accuracy and efficiency of simulations but also help create digital twins for different industries. As emerging technologies continue to rise, the acceptance of DT is projected to grow rapidly, promoting greater innovation and transformative change across various industries. The future of digital twins looks promising, with ongoing advancements setting the stage for more integrated, interactive, and intelligent systems.

**Keywords** The Internet of Things (IoT) and Artificial Intelligence (AI) · Digital twin · 5G · BIM · GIS · Augmented reality/virtual reality · Cloud computing

P. K. Garg (✉)
Indian Institute of Technology, Roorkee, India
e-mail: pkgiitr@gmail.com

N. K. Tripathi
School of Engineering and Technology, Asian Institute of Technology, Klong Luang, Pathumthani, Thailand
e-mail: editor@e-geoinfo.com



L. Sharma et al. (eds.), *Digital Twin Technology in Geospatial Science: Trends, Applications and Challenges*, Data-Intensive Research,
https://doi.org/10.1007/978-3-032-20311-3_2

## 1 Introduction

A DT is a virtual replica that integrates digital technologies to create a real-time representation of physical assets like an object, system, or process and their operational states. It is not just a 3D model or a simulation but a virtual representation of physical assets that uses data from its physical counterpart. Various sensors collect and provide continuous data from their physical counterpart that is used for analysis, simulation, and modeling of an object. DT involves visualization of a real object, which allows interaction, monitoring, modeling for the benefit of its physical counterpart, system, or process.

Figure 1 presents a digital twin architecture that has three key elements: (i) a physical world, (ii) a virtual world, and (iii) an interconnectivity between them. Each element further integrates other components, depending on the need. However, some basic components are; sensors in the real-world (to collect data or information), databases a physical object, edge processing software, data security, DT, data processing capabilities supported by ML, AI, big data, etc., and communication devices, like internet, Bluetooth, satellite, etc. The digital twin architecture provides data visualization for the users. Both the digital twin and physical object are connected through communication devices with protocols and interfaces, like WiFi, Bluetooth, and wired connections. The architecture would allow continuous monitoring and imagining [1].

The DTs are very useful in various industries, particularly in manufacturing industries. Outside manufacturing applications, the digital twins have been used for 2D/

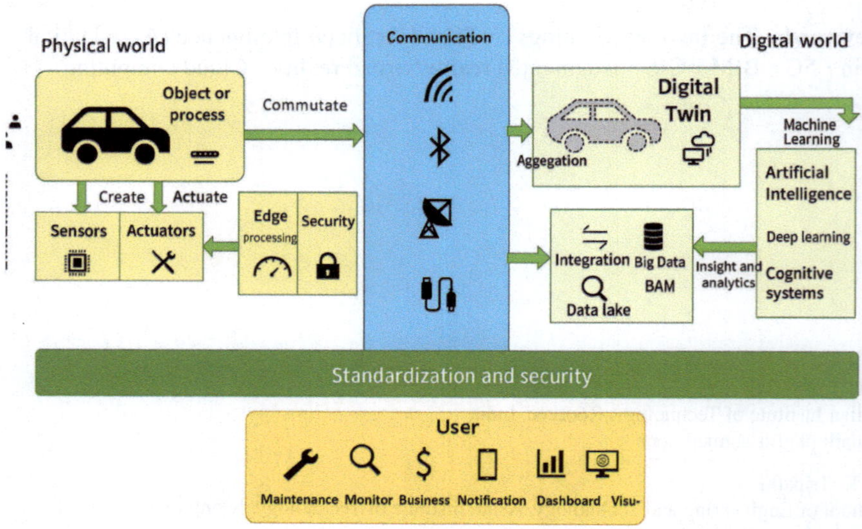


**Fig. 1** A digital twin architecture

3D models to Building Information Modeling (BIM) and the Metaverse (Virtual/Augmented Reality) [2]. The role of geospatial science (GS) within DT is demonstrated in various applications, particularly in the urban sector, where 3D geospatial data are used to represent physical assests in a virtual mode, such as buildings, structures, and their surrounding environments. The urban digital twins have emerged as 3D city models, BIM, and smart cities. Digital twins bring innovation and enhance knowledge sharing across industries, governments, and communities.

Geospatial science is a field that uses information technology to understand and analyze the Earth and human societies, focusing on the spatial relationships and patterns of geographic data. It involves collecting, managing, analyzing, and visualizing geospatial data using technologies, such as (i) remote sensing, (ii) Geographic Information Systems (GIS), (iii) Global Positioning Systems (GPS), (iv) Unmanned Aerial Vehicles (UAVs), and (v) Light Detection and Ranging (LiDAR). Geospatial science has been referred to in the context of digital twins in a no. of ways, for example, location is an important parameter that allows the objects to be visualized on a map. In addition, the geometric structure may have a geospatial element, mainly in urban digital twins, where the element may be a building with a specific shape [3–6].

## 2 Geospatial Digital Twins

Geospatial data serves as a very crucial foundation for data based decision-making across various sectors such as urban planning and development, environmental management, disaster mitigation, and infrastructure development. It is used to establish spatial relationships, patterns, and changes. A digital twin connected to geographic locations becomes "geospatial," that allows a real-time visualization of physical assets. By integrating technologies, like IoT, AI, GPS, and GIS, geospatial digital twins help accurate decision-making [7].

Geospatial digital twin is defined as "serving as a mechanism for monitoring, visualizing, analyzing, optimizing, and forecasting the behaviors and processes linked with corresponding physical systems." A geospatial DT is much beyond creating a simple 3D model. It represents a virtual environment that combines physical objects, processes, and relationships. Geospatial digital twins provide a location-based representation of objects, infrastructure, and systems, along with their sizes. These virtual models combine current data, 3D visualization, and advanced analytics to offer new insights and open new doors for innovation.

The geospatial digital twins can cover buildings, other infrastructure, transportation networks, cities, etc. They are frequently utilized for urban planning, lifecycle management, and the continuous monitoring of smart city operations but can also be applied in agriculture, energy, natural resources management, transport, health, environment, etc.). Geospatial DT are changing the way we apply geospatial data in various applications, ranging from smart cities to resource monitoring.

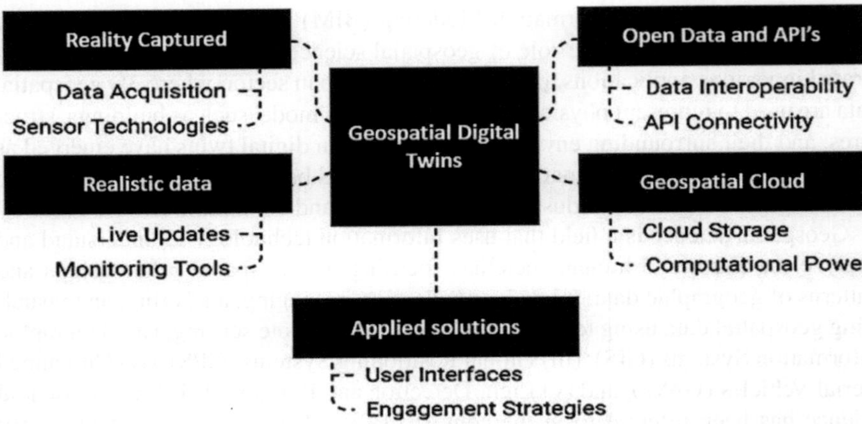


**Fig. 2** Five core elements of a geospatial digital twin

There are five core elements to lay the foundations of a geospatial DT, as shown in Fig. 2. Reality capture requires the use of equipment and sensors for mapping and modeling, and real-time data is collected for continuous monitoring and updates. The digital twins can work effectively utilizing 3D and 4D temporal data, which requires dimensional accuracy and precise location for the representation of assets, infrastructure, and systems. Open data & APIs are integration frameworks for connecting systems. The geospatial cloud consists of infrastructure for scalability. Applied solutions consist of user interfaces for faster decision-making. Integrating digital twin technologies into GIS allows geospatial analysis to provide a dynamic approach for geospatial modeling and 3D visualization. They also offer simulation capabilities for improved understanding of complex spatial dynamics. GIS-based 4D digital twins represent changes through historic, current, and future states.

Presently, the use of GIS and digital twins in various disciplines based on geospatial data is increasing, including infrastructure planning, disaster response, urban development, water resources, and environmental management. Geospatial analysis in digital twins can streamline processes, strengthen insight, and help in decision-making. GISs are vital to creating and maintaining digital twins due to their capabilities to integrate different types of spatial data at a single platform, visualize data to create interactive maps, identify patterns and relationships between objects through spatial analytics, and real-time monitoring of assets.

## 3 Emergent Technologies for Digital Twins

Several innovative tools and technologies have given rise to digital twin technology and enhanced its capabilities. These technologies enable a complete, real-time perspective of an object, operations, and potential scenarios through an integrated

DT. They allow the buildup of dynamic, interactive visualizations by aggregating huge amount of data from business systems, sensors, and information models. The fully detailed DT models can simulate the effects of plans and concepts on persons or individuals or places within a virtual, risk-free environment. In fact, DT are part of everyday life, often without our awareness. For instance, Google Maps functions as a DT of the transportation network, providing all possible routes along with up-to-date information on traffic, construction, and other conditions [8–11].

The DT require a vast amount of real-time data, called big data. Big data is to be processed and analyzed, converting this data into actionable information. The machine learning algorithms further boost the analysis and simulate various scenarios, as machine learning algorithms learn from these simulations to enhance the forecasts and decision-making. The fusion of digital twins with AR provides opportunities for users to interact with virtual replicas in a real-world environment. The computational power of quantum computing would further enhance the capabilities of DT, make simulations more complicated and precise predictions [12].

Figure 3 presents six main stages required to develop a fully operational digital twin [13]: (i) a virtual replica of outer world, (ii) data processing to guide design decisions, (iii) simulating the product's behavior within a virtual environment, (iv) designing of physical product to achieve the desired performance, (v) establishment of secure, real-time, two way data communication between real and virtual systems, and (vi) collection of all product-related data from different sources. Several emerging technologies are critical for implementing these steps, some of which are outlined below.

## *3.1 Internet of Things (IoT)*

It is a network of devices that are interconnected via the internet. It enables everyday objects often called "things" to become intelligent by allowing them to sense, collect, and share data about their surroundings [14]. IoT serves as a basic component of DT systems, as these systems depend on extensive and continuous data gathering. IoT technologies encompass a range of elements (sensors, actuators, software, and electronic components). The combination of IoT has various advantages, such as improved automation, precision, operational efficiency, and overall productivity [13]. By applying IoT, data can be collected on real time basis from physical objects, these objects or physical entities can communicate and collaborate autonomously, IoT-connected products can offer better online services, and physical systems can be remotely observed, managed, and adjusted.

The IoT sensors gather all the information and deliver it to the cloud that can be used by other systems. The IoT is a primary driver for DT technology, as IoT-based devices collect real-time data from their physical environment, which is essential for the successful creation and implementation of a DT of the physical object. This continuous time-series data can help in monitoring, maintenance, and operational planning of an equipment/a machine. With the help of IoT sensors, we can keep

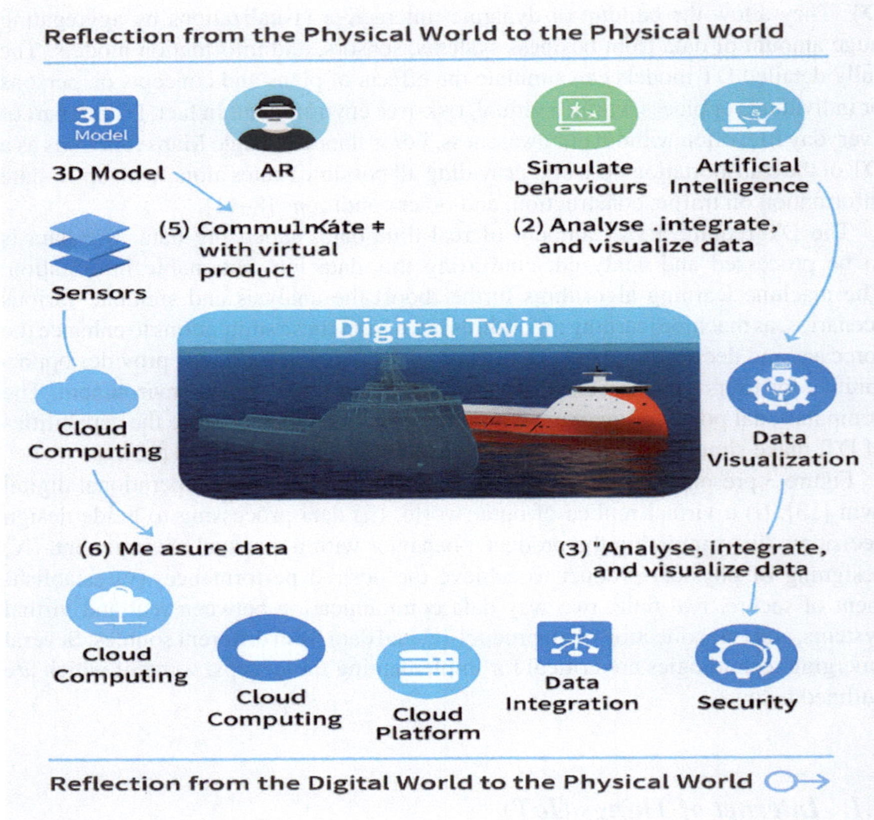


**Fig. 3** Steps used to create a digital twin employing emergent technologies

track of equipment's performance, which is fed into predictive analytics to forecast potential failures before it actually occurs.

## 3.2 5G and Advanced Network Infrastructure

The 5G networks are a boon for digital twins, particularly where real-time data transfer and connectivity are critical. It provides much lower latency and higher bandwidth, enhancing the responsiveness and accuracy of digital twins. Applications, like autonomous driving, remote surgery, and real-time remote monitoring of industrial equipment, essentially require 5G support as even small delays can lead to adverse results.

### 3.3 Global Positioning System (GPS)

GPS is referred to as an enabling technology and tool for data transmission and acquisition. The map is referred to as an important technology for visualization. Geospatial data from GPS devices has enabled GIS platforms to create, render, and analyze infrastructure data at a temporal scale. GPS technology plays a vital role in DT by providing real-time location data, which is essential for creating accurate and dynamic virtual copy of physical system. This data allows digital twins to mirror the location and movement of real-world objects, required for applications, such as smart cities, autonomous navigation, and resource management.

### 3.4 Big Data Analytics

With the rapid growth of advanced technologies like IoT and AI, data has become a key component of advanced technological innovation. It is generated and collected from various sources, such as social interactions, machine operations, and transactional activities. When the scale, diversity, and velocity of this data reach significant scales, it is collectively referred to as *Big Data*. The real-time applications of Big Data are wide-ranging and transformative about predictive analytics, consumer behavior analysis, operational efficiency enhancement, demand forecasting, real-time system visualization, and the development of personalized marketing strategies.

DT systems generate and process large volumes of data collected from multiple sources. Within the manufacturing domain, Big Data includes large quantities of structured, semi-structured, and unstructured information which is produced throughout the complete product lifecycle. Most of these data are captured automatically and in real time through IoT-enabled devices and sensors. By employing Big Data analytics supported by cloud computing infrastructure, manufacturers can identify production bottlenecks, determine their underlying causes, and formulate effective solutions. Manufacturing operations can also enhance their efficiency, strengthen competitiveness, and facilitate the transition toward smart manufacturing environments. Recently, both technologies have gained significant attention and are widely recognized as foundational pillars of smart manufacturing. A comparative evaluation of these two technologies across various dimensions is illustrated in Table 1 [15].

Big data analytics technology is employed to process and analyze this data to derive useful results. Techniques, like data mining, ML, and statistical analysis, are used to identify patterns and correlations within the data, which help in optimizing processes, predicting maintenance needs, and improving overall performance. Big Data, however, presents many challenges, such as managing data from various sources, quality issues, dealing with integration complexities, and managing the escalating costs of computing resources.

**Table 1** A comparison of big data and digital twin [15]

| Items | Big data | Digital twin |
|---|---|---|
| Background | Wide applications of ICT and the exponential growth of data have raised the requirement of Big Data | The rapid development of ICT and need for cyber-physical integration have raised the requirement of Digital Twin |
| Concept | Includes large volume of data and the processing | Includes the data of physical and virtual objects |
| Functions | Mining behavior and patterns of features; Establish the trends; Data visualization; Predict and analyze problems; Help decision making; Optimize and improve process | Virtual verification; Simulation running; Real-time monitoring; Predict and diagnose the problems; Optimize and improve the process |
| Applications | Product lifecycle from design to maintenance, repair, etc. | Product lifecycle from design to maintenance, repair, etc. |
| Effects | Improve efficiency, customer satisfaction and precision; Extend life of product; Reduce the cost; Promote smart manufacturing | Improve efficiency, customer satisfaction, and precision; Extend the life of product; Reduce the cost and development cycle; Promote smart manufacturing |
| Key technology | IoT, Cloud computing, Fog computing, Data cleaning, Data mining, ML, etc. | IoT, Data analysis, Virtual reality, Augmented reality, CPS, Simulation, etc. |
| Data sources | From the physical entities, information systems and Internet | From the physical entities, virtual models, and their fusion |
| Data volume | Large, ranging from PB to EB, even ZB | No specific quantity |
| Data features | Structured, semi-structured and unstructured data | Structured, semi-structured, and unstructured data |
| Multisource correlation | Focus on data attributes and highlight the relationships between features | Focus on the consistency of multi-source data and their evolution and integration |
| Data acquisition | Sensors, RFID, and other sensing devices; SDK, API; Web crawler, etc. | Sensors, RFID, and other sensing devices; Model data interfaces, etc. |
| Data process -ing | Through processing tools, algorithms platforms, etc. | No specific methods |
| Data fusion | Data fusion of various objects in single phase of the product lifecycle | All elements: whole process, whole business data fusion in the entire product lifecycle |
| Visualization | Table, chart, graphs, prints, etc. | Image, video, virtual and augmented reality, etc. |
| Result verification | Through the physical execution process, or the simulation from the third party; Relatively slow | Through its own virtual simulation and evolution functions to execute pre-verification; Relatively advanced |

## 3.5 Artificial Intelligence and Machine Learning

Artificial Intelligence (AI) and Machine Learning (ML) are equally important in promoting the use of digital twin technology. The AI is the first topic of interest in data analytics. It helps analyse the gathered data automatically, predict possible outcomes, and provide suggestions to minimise potential problems. When working with a digital twin, AI can enhance value, for example, AI-based control algorithms can be trained with the digital twin for uninterrupted production [14]. Generative AI could also enhance decision-making by providing data-driven insights and simulations that predict the outcomes of various design and construction scenarios with precision. Through predictive analytics and scenario simulations, AEC professionals would be able to evaluate project feasibility, foresee potential challenges, and make well-informed decisions to mitigate risks [16].

The ML algorithm, a subsection of AI, provides computer the ability to learn and act for the users. The ML uses sophisticated algorithms to collect and analyse data autonomously. Digital twins use AI and ML algorithms for predictive learning, for example, virtually real-time tracking the movement of materials through a supply chain with help from IoT sensors. With AI and ML, that same system can now support process improvements. The AI algorithms can be employed to analyze a large amount of data generated by the IoT devices, and identify patterns and detect anomalies efficiently and accurately. The ML algorithms can learn from these huge datasets and make accurate predictions for proactive maintenance, optimizing operations, and enhancing decision-making processes in real-time. As digital twins becomes more productive by leveraging advanced technologies, like AI, machine learning, and predictive analytics.

## 3.6 Augmented Reality (AR) and Virtual Reality (VR)

Augmented Reality (AR) and Virtual Reality (VR) technologies can improve the human–machine interaction by combining the real and virtual world, and allow decision-makers to visualize complex data and simulations more deeply, thus enhancing the capability of digital twin. The AR can overlay digital twin data onto a physical object to provide a comprehensive view of potential issues and improvements. The VR can create a fully immersive environment to interact with the digital twin as if it were a real object. This is extremely useful to manipulate complex systems in a controlled and risk-free virtual environment.

## 3.7 Blockchain

The Blockchain has become the next step in redefining the concept of digital twins. Blockchain technology enhances the applications of digital twins due to its decentralized and immutable nature. In blockchain technology, a digital ledger ensures data integrity, and allows efficient and safe information exchange by encrypting data through blocks using Hash algorithm and securely recording transactions across multiple nodes using cryptographic methods. For the implementation of digital twins, several challenges, such as data interoperability, privacy concerns, and property rights issues, are faced. However, the integration of blockchain technology into digital twins overcomes these challenges, and enhance the potential for advanced implementations.

It is an essential tool to enhance the application of digital twins in IoT, transferring the data and value onto the Internet with full transparency. Blockchain technology brings a dimension of security and transparency to data. Security of the data and digital twin platform is critical. By integrating blockchain with digital twins, every change, decision, and data can be recorded in an immutable ledger, enhancing the reliability and traceability. It is a way to generate and monitor digital twins due to the security and immutability of technology. Sectors, such as manufacturing and retail are set to benefit from these technologies interlinking, connecting the physical with the digital world.

Digital twins of an item with a blockchain will help with brand protection, keep its transaction records, along with information about its production, and its movement. The blockchain with digital twins will allow companies tracking the products globally with greater accuracy, eliminating the chances of stolen goods across the globe [17]. Blockchain technology has the potential to become the most relevant and capable way to generate and monitor digital twins due to security reasons. The key features of this technology include distributed ledger, encryption algorithm, tracking algorithm, and data reliability, which are coupled together for various digital twin applications. These include organizational information management, information and data security, information verification, and the enhancement of information systems within digital twins. Specialized distributed ledger platforms that allow sharing information amongst digital twins would be required.

There are several application areas for future research on complex and in-depth integration of blockchain technology into digital twins, as per framework given in Fig. 4. It involves exploring advanced methodologies, such as AI and big data, which can focus on improving affordable technology and updating data processing methods with innovative techniques [18].

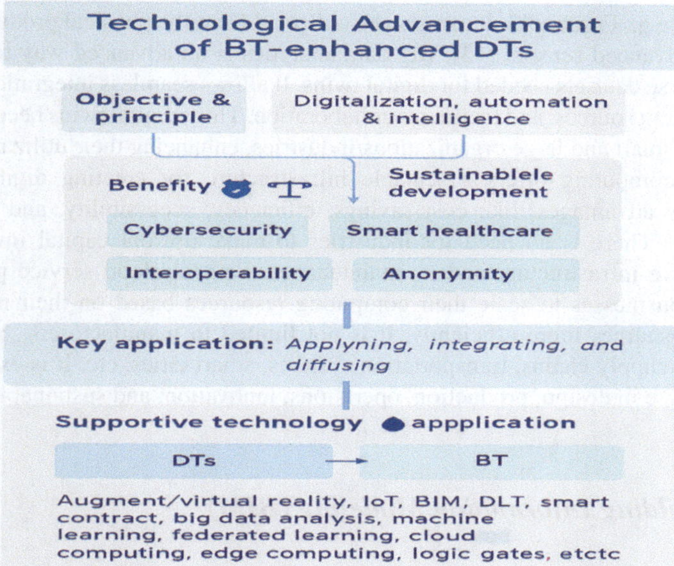


**Fig. 4** Future development framework of integration of blockchain technology for creating digital twins [18]

## *3.8 Cloud Computing*

Cloud computing refers to the provision of computing resources, infrastructure, software, data and information management services and applications via the Internet. The benefits of cloud-based computing include easy access to computing resources and the flexibility to scale systems with business needs. Cloud services may be delivered in different ways in terms of their ownership, structures, and operations [19]. For example, the public cloud is owned and managed by the service providers that offer a selection of computing resources (e.g., Amazon Web Services, Microsoft Azure), and the resources can be accessed over the Internet for external use. Whereas, the private cloud is self-operated and available for internal use only.

The applications of digital twins require huge data, and cloud computing allows for storing all data in the virtual cloud and easily accessing the required info from any place. Cloud computing provides the necessary infrastructure support for digital twins, such as large storage and powerful computational capabilities required for handling the large datasets and complex simulations performed by digital twins. Big data contains valuable information that are characterised by high volume, velocity and variety. Big data analytics aims to extract useful information from an enormous amount of highly unstructured and seemingly unrelated data. Big Data may be available on the Internet, as it is frequently generated by customers using Internet, for example, customer reviews on products (e.g., Amazon) and lots of information social platforms (e.g., Facebook). It can also be collected from the physical products using IoT and advanced sensors [13]. Big data analytics is an advanced way to analyse huge, diverse datasets needed for digital twins. It allows seamless integration of data from different sources, and facilitates collaboration. Thus, digital twins become more usable for small and large organizations/industries, enhancing their utilization.

Cloud computing offers a scalable infrastructure for creating digital twins. It provides advantages, like cost savings, efficiency, accessibility, and powerful computing. There is no need for industries to make upfront capital investments in expensive infrastructure and its maintenance. Several cloud service providers, allowing businesses to scale their computing resources based on their needs and allocate resources more efficiently. It is not limited to manufacturers, but it also extends to supply chains, transportation systems, smart cities, etc. It is expected to revolutionize in design, production, operations, innovation, and sustainability.

## 3.9 Building Information Modeling (BIM)

In the field of architecture and construction, the latest tools and technologies have created digital revolution. The Building Information Modeling (BIM) is one of them, having evolved from a simple 3D model to sophisticated digital twins. While BIM focuses on the design and construction phase, digital twins come into play after completion. They capture real-time data from IoT sensors, connected devices and tracking systems, integrating it into the original BIM model. This integration allows a better understanding of asset performance and condition.

The BIM has reshaped how we design, construct, and manage buildings with efficiency and precision. BIM offers a static representation of the project, while digital twins provide a dynamic and contextualized visualization of assets. This means users can explore and interact with digital models in a virtual environment, enabling a deeper understanding of how assets behave in their real-world context. BIM introduced digital 3D models representing a building's physical and functional characteristics, which helps improve efficiency and reduce errors during the construction phase and throughout the building's lifecycle.

The IoT provides real-time monitoring and continuous data on factors, such as temperature, humidity, and equipment status of BIM. With the integration of simulation tools, engineers and architects could perform structural analyses, energy simulations, and other critical assessments directly within the BIM environment. It allowed for more informed decision-making, optimized designs, and ultimately improved building performance. Using advanced techniques like laser scanning and point cloud data, it became possible to create accurate BIM models reflecting the real-world state of a building. This innovation revolutionized renovations, retrofits, and maintenance, facilitating greater precision and efficiency. The use of AI allowed predicting the maintenance needs, energy consumption patterns, and even design issues. This data-driven decision-making process empowered stakeholders throughout a building's lifecycle.

In the construction industry, BIM technology along with digital twins creates very detailed and dynamic models of buildings and infrastructures. It facilitates the visualization, simulation, and analysis of construction projects over their entire lifecycle. Integrating BIM with digital twins enhances the ability to real-time monitor construction progress, manage changes, and ensure compliance with design specifications.

## 3.10 Geographic Information Systems (GIS)

Accurate representations of complex systems require robust maps and spatial analytics from GIS technology. GIS-based digital twins go beyond 3D models as they reflect change over time, showing historic, current, and future states. GISs are powerful spatial analysis tools that enhance the understanding of geographic data for the digital twin. The integration of GIS allows for simulations of environmental impacts, urban growth, and infrastructure resilience, providing planners and decision-makers to make informed decisions. From single facility to large built and natural systems, geospatial digital twins scale to meet the changing needs.

GIS and related geospatial technologies are essential to creating geospatial digital twins, providing locational data needed to map a digital replica to the corresponding physical space. GIS unifies different types of spatial data—topography, infrastructure, environmental, and land use, into a single platform hosting a detailed digital twin. Data obtained from various sources is used in GIS to build an interactive map representing a cohesive view of all objects and processes in physical space. GIS systems can incorporate real-time data provided by IoT devices and update corresponding assets in the digital twin, ensuring synchronization of real and virtual conditions. Location data obtained from GIS becomes the basis for identifying patterns and relationships between spatial objects, as well as for generating insights about their conditions. Based on GIS data, digital twins can model the effects of both natural and technological changes, running various scenarios and predicting their outcomes. GIS enhances decision-making opportunities by placing objects, events, and their relationships in spatial context, highlighting trends and patterns, which enable accurate predictive analytics and data-driven decision-making.

These technologies are used along with a digital twin to create a 3D model of the physical object, geographic mapping, and real-life processes. The geospatial data transforms the 3D model into a multi-dimensional model of the digital twin, with all live and historical information. A continuous stream of data from IoT devices will update the digital twins to match the events and changes affecting its real object. Digital twins with AR/VR may accelerate the learning capabilities. Digital twins with AI and ML can be used to improve real-time working functions and reduce errors, as digital twins will take input data from various sources.

Digital twins enable us to transform many of our business processes, including project engineering design, construction execution planning, supply chain and materials handling, and facility operations and maintenance. The speed of business with digital twin technology is increasing at an alarming rate. The innovative technologies today, like AI, ML and IoT, can be used to build a virtual framework that coordinates complex events across space and time.

Despite the available technologies, there are several challenges of data-driven product design [13]. It includes effective transformation of Big Data to useful information that can be directly derived to support the design decision-making process. Another challenge is to integrate different data sets, which are collected from both Internet and IoT using various methods, in order to discover the patterns. Rapidly responding to a current event on real-time data is another challenge, including the prediction based on historical data.

## 4 Conclusion

Digital twins have simplified several things for companies and manufacturers. A digital twin uses real-time data to lower the costs, enhance productivity, predict maintenance requirements, and increase overall efficiency. Real-time data empowers decision-makers to make informed choices. Optimized supply chains can minimize the wastage. Firms can use the digital twin technology to increase profit, remove errors, improve product quality, and automate the process. Digital twin technology can harness real-time data to drive learning and improve overall performance. Bringing together hardware, software, and data, digital twins are useful to optimize the design and operation of a process in real time.

Developing digital twins requires continuous data for ongoing operations, which empowers networked software systems with the resources to accelerate decision-making. The AI, ML, and other advanced technologies can be used not only to suggest improvements in operations but also to independently "learn" and improve performance with time. It requires integrating past historical data along with machine learning algorithms to monitor performance in real-time [20].

The true power of digital twins is to replicate real-time objects/data. The data provides a solid foundation for predictive analytics. Despite several associated technological advancements, there is an immense potential of using digital twins with GIS. This integration can be used enhance decision-making, improve operational efficiency, and lower the costs. Overall, the future of integrated digital twins and GIS is promising.

Digital twins will continue to shape industrial sector and other sectors in the future. Organizations may utilise these digital twins for various tasks, such as optimizing energy consumption, reducing carbon emissions, and enhancing efficiency. From designing eco-friendly products to sustainable urban planning, digital twins will prove to be a useful tool in achieving environmental goals to create a greener and more sustainable future.

## References

1. Botín-Sanabria DM, Mihaita A-S, Peimbert-García RE, Ramírez-Moreno MA, Ramírez-Mendoza RA, Lozoya-Santos JDJ (2022) Digital twin technology challenges and applications: a comprehensive review. Remote Sens 14:1335. https://doi.org/10.3390/rs14061335
2. WGIC Policy Report: 2022–01, Spatial digital twins: global status, opportunities, and the way forward. https://wgicouncil.org/wp-content/uploads/2023/11/WGIC-Policy-Report-2022-01-Spatial-Digital-Twins.pdf
3. Sharma L, Garg PK (eds) (2025) Digital twins and simulation technology concepts and applications, 1st edn. Chapman and Hall/CRC. https://doi.org/10.1201/9781003582489
4. Sharma L, Garg PK (eds) (2024) Deep learning in Internet of Things for next generation healthcare, 1st edn. Chapman and Hall/CRC. https://doi.org/10.1201/9781003451846
5. Sharma L, Garg PK (eds) (2023) Technological prospects and social applications of society 5.0, 1st edn. Chapman and Hall/CRC. https://doi.org/10.1201/9781003324720
6. Sharma L, Carpenter M (eds) (2022) Computer vision and Internet of Things: technologies and applications, 1st edn. Chapman and Hall/CRC. https://doi.org/10.1201/9781003244165
7. Ranatunga S, Ødegård RS, Onstein E, Jetlund K (2024) Digital twins for geospatial decision-making. In: ISPRS annals of the photogrammetry, remote sensing and spatial information sciences, vol X-4–2024, ISPRS TC IV Mid-term Symposium "Spatial Information to Empower the Metaverse", 22–25 October 2024, Fremantle, Perth, Australia
8. Sharma L, Garg PK (eds) (2021) Artificial intelligence: technologies, applications, and challenges, 1st edn. Chapman and Hall/CRC. https://doi.org/10.1201/9781003140351
9. Sharma L (ed) (2020) Towards smart world: homes to cities using Internet of Things, 1st ed. Chapman and Hall/CRC. https://doi.org/10.1201/9781003056751
10. Sharma L, Garg PK (eds) (2019) From visual surveillance to Internet of Things: technology and applications, 1st edn. Chapman and Hall/CRC. https://doi.org/10.1201/9780429297922
11. Sharma L (2019) Object detection with background subtraction. LAP LAMBERT Academic Publishing, SIA OmniScriptum Publishing, Latvia, European Union, February 2019. ISBN: 978-613-7-34386-9
12. Ranjan S (2023) Industry 4.0 and digital twin technology: a symbiotic relationship, October 23, 2023. https://www.toobler.com/blog/industry-4-0-and-digital-twin
13. Tao F, Sui F, Liu A, Qi Q, Zhang M, Song B, Guo Z, Stephen C Lu, Y, Nee AYC (2018) Digital twin-driven product design framework. Int J Prod Res. https://doi.org/10.1080/00207543.2018.1443229
14. Fuller A, Fan Z, Day C, Barlow C (2020) Digital twin: enabling technologies, challenges and open research. IEEE Access, 1–1. https://doi.org/10.1109/ACCESS.2020.2998358
15. Qi Q, Tao F (2018) Digital twin and big data towards smart manufacturing and industry 4.0: 360 degree comparison. IEEE Access 6:3585–3593. https://doi.org/10.1109/ACCESS.2018.2793265
16. Spiegel R (2024) Digital twins have become a blend of multiple technologies, October 31. https://www.designnews.com/automation/digital-twins-have-become-a-blend-of-multiple-technologies
17. Miskinis C (2018) How the blockchain can improve the concept of digital twins, March 2018. https://www.challenge.org/insights/blockchain-and-digital-twin/
18. Ma X, Du W, Hongping Y, Li L, Jing L (2024) Examining the nexus of blockchain technology and digital twins: bibliometric evidence and research trends. Front Eng Manag 11:481–500. https://doi.org/10.1007/s42524-024-0306-4
19. Zhang G, MacCarthy B, Ivanov D (2022) The cloud, platforms, and digital twins—enablers of the digital supply chain. https://doi.org/10.1016/B978-0-323-91614-1.00005-8
20. Mathieu A (2020) 6 ways next generation digital twins will transform business, 24 February 2020. https://vantiq.com/blog/digital-twins-will-transform-business/

# Chapter 3
# Mirroring the World: Unleashing the Power of Advancements in 3D Modeling and Visualization in GIS Digital Twins

**Naman Agrawal**

**Abstract** Geographic Information System (GIS) digital twins are revolutionizing spatial governance by creating dynamic, real-time 3D virtual representations of physical environments. Rapid advances in 3D modeling, data acquisition via LiDAR and drones, and visualization platforms like Esri's ArcGIS and Unreal Engine empower policymakers to simulate scenarios improve urban resilience, optimize infrastructure, and enhance environmental management. Leading cities such as Boston, Helsinki, and Singapore showcase transformative impacts—with Boston reducing planning errors by 18%, Helsinki covering 98% of its building stock, and Singapore boosting airport efficiency by 12%. However, challenges remain, including data governance complexities, a widening urban–rural digital divide where only 10% of projects serve rural areas, and resource constraints highlighted by a shortage of 200,000 skilled geospatial professionals globally in 2025. The geospatial digital twin market is poised to surge to $25 billion by 2026 and $110 billion by 2028, underscoring the critical need for equitable policy frameworks, interoperability standards, and capacity-building initiatives. This chapter synthesizes technological innovations, global adoption trends, and policy implications, offering actionable recommendations for fostering inclusive, sustainable governance through GIS digital twins. It calls for balanced investment, collaborative innovation, and ethical stewardship to harness this powerful tool in addressing urbanization, climate change, and infrastructure resilience for a smarter, safer future.

**Keywords** GIS digital twins · 3D modeling · Urban resilience · LiDAR · Environmental management · Real-time data · Interoperability · Digital divide · Geospatial analytics · Policy frameworks

N. Agrawal (✉)
Science, Technology and Innovation, Embassy of the Kingdom of the Netherlands, New Delhi, India
e-mail: naman.niti@gmail.com



L. Sharma et al. (eds.), *Digital Twin Technology in Geospatial Science: Trends, Applications and Challenges*, Data-Intensive Research,
https://doi.org/10.1007/978-3-032-20311-3_3

# 1 Introduction

Geospatial technologies have rapidly evolved, ushering in a new era of spatial analysis and governance with Geographic Information System (GIS) digital twins at the forefront of this transformation. Defined as dynamic, real-time virtual representations of physical environments, GIS digital twins integrate advanced three-dimensional modeling and visualization techniques. This allows them to transcend the limitations of traditional two-dimensional mapping. Policymakers, urban planners, and environmental managers can now interact with spatial data in a multidimensional manner through GIS digital twins. This offers a new lens to address pressing global issues like rapid urbanization, climate change, and improving infrastructure resilience. Technological advancements are reshaping how these stakeholders work with spatial data to solve complex problems. GIS digital twins provide insights that can inform policy and planning over diverse geographic areas. Their dynamic nature also enables monitoring changing conditions in real-time, allowing for more agile responses to emerging issues across diverse landscapes and communities.

The explosive progression of geospatial technologies has ushered in a new era defined by intricate spatial examination and administration, with Geospatial Information Systems (GIS) digital twins situated at the vanguard of this transformation. Described as dynamic, real-time cybernetic representations of material environments, GIS digital twins integrate sophisticated 3D rendering and visualization methods to surmount the restrictions of conventional two-dimensional mapping [1–4]. This segment explores how these technological enhancements are reworking the means by which policymakers, urban planners, and environmental administrators interact with spatial information, offering a multidimensional lens through which to tackle urgent worldwide tests like urbanization, environmental change, and infrastructure resilience. Furthermore, GIS digital twins supply decision makers with a spatiotemporal perspective that facilitates evidence-based planning for challenges such as integrated transportation networks, pollution remediation, and preemptive disaster response.

At the heart of this chapter lies a critical examination of the technological advancements propelling GIS digital twins forward. Innovations in 3D modeling—such as procedural generation and semantic integration—combined with cutting-edge visualization platforms like Esri's ArcGIS and game engines such as Unreal Engine, have elevated the precision and accessibility of geospatial twins [5], 6. These innovations are not simply technological breakthroughs,they hold profound implications for public policy, enabling governments to simulate policy effects, make more informed decisions about resources, and engage citizens in planning. For instance, the ability to visualize flood situations in real-time or anticipate trends in urban sprawl empowers governments to craft forward-looking, rather than reactive, responses—a shift that can redefine twenty-first-century governance [7, 8].

But the arrival of GIS digital twins comes with problems. The confluence of vast datasets from data sources like Light Detection and Ranging (LiDAR), photogrammetry, and the Internet of Things (IoT)—with over 31 billion devices globally in

2025 [9]—poses problems of data governance, privacy, and equity. Moreover, as urban use cases flourish, rural regions trail behind, symbolizing a digital divide that policymakers will have to bridge [10]. Chapter three tries to resolve these dynamics in a balanced assessment of the future prospects and pitfalls. It posits three foundational questions: What have recent innovations in 3D modeling and visualization contributed toward GIS digital twins? What are tangible applications converging in worldwide public administration? And what kind of policy structure needs to be established to ensure their equitable and sustainable implementation? By combining technical expertise, global trends, and exemplars, this analysis aspires to bring to the fore the change potential of GIS digital twins and guide their integration into modern governance.

## *1.1 Global Adoption Trends*

North America leads the charge, with the United States at the forefront, driven by robust technological infrastructure and significant public investment. Boston's digital twin, for example, integrates real-time data on zoning, transportation, and public safety, reducing planning errors by 18% in 2024 trials [5]. This success has spurred initiatives in cities like New York and Los Angeles, where digital twins are being piloted to optimize traffic flows and enhance disaster preparedness. The U.S. geospatial market alone is projected to account for 40% of the global $25 billion valuation by 2026, a testament to its early adoption and innovation [11].

Europe follows closely, with a focus on sustainability and regional collaboration. The European Commission's Destination Earth initiative, launched in 2022 and expanded in 2024, exemplifies this trend, using GIS digital twins to model climate scenarios and inform environmental policy across member states [12]. Cities like Helsinki and Amsterdam have embraced 3D visualization for urban planning, with Helsinki's digital twin covering 98% of its building stock by 2025, aiding in energy efficiency policies [13]. The European market is expected to grow at a compound annual growth rate (CAGR) of 15% through 2030, fueled by stringent environmental regulations and cross-border data-sharing frameworks [11].

In Asia, rapid urbanization and infrastructure development drive adoption, particularly in Singapore and China. Singapore's Changi Airport digital twin, updated in 2024, enhances operational efficiency by 12% through passenger flow simulations, setting a benchmark for transport hubs globally [14]. Meanwhile, China's smart city initiatives, such as Shenzhen's digital twin, leverage 3D GIS to manage population densities exceeding 6000 per $km^2$, integrating IoT and BIM for real-time urban governance [15]. Asia's share of the geospatial twin market is projected to rise from 20% in 2025 to 30% by 2028, reflecting its aggressive technological uptake [12].

The Global South, while lagging, shows promising growth. In Africa, South Africa's Cape Town has piloted a GIS digital twin to monitor water resources, addressing drought challenges with 3D visualizations of catchment areas [16]. Latin America's adoption is led by Brazil, where São Paulo's digital twin supports flood risk management, though only 10% of projects in these regions target rural areas, underscoring a persistent urban bias [10]. The Middle East, particularly the United Arab Emirates, is advancing through projects like Dubai's Virtual Dubai, which models the entire emirate in 3D to support its 2040 Urban Master Plan [11].

Globally, adoption is uneven, with 70% of smart city projects incorporating BIM-GIS workflows in 2025, up from 45% in 2020 [15]. Yet, rural applications remain limited, comprising just 10% of initiatives, highlighting a need for scalable, cost-effective solutions like drone-based mapping [10]. Investment trends suggest a shift, with $5–10 million per city-scale twin becoming standard, though public–private partnerships are increasingly vital to offset costs [11]. As these technologies mature, their global spread will depend on policy innovation, balancing technological ambition with equitable access.

## 2 Technological Advancements

The evolution of GIS digital twins hinges on remarkable strides in 3D modeling and visualization technologies, which have transformed static geospatial data into dynamic, real-time representations of physical environments [17, 18].

These developments—ranging from data gathering, modeling techniques, and visualization tools—empower policymakers and practitioners to model complex scenarios, monitor systems with precision, and negotiate stakeholders at maximum efficiency. This section delves into the most impactful developments behind this revolution, drawing on emerging trends and empirical examples to illustrate their impact. With the geospatial digital twin market exploding toward an estimated $25 billion valuation by 2026 [11], it is worth having an understanding of these technological underpinnings in order to be able to take advantage of their potential in public policy and beyond.

### *2.1 Data Acquisition and Integration*

The foundation of any GIS digital twin lies in its ability to acquire and integrate high-fidelity spatial data. Light Detection and Ranging (LiDAR) has emerged as a pillar technology with resolutions of only 1 cm, which facilitates high-level 3D reconstructions of cities and natural scenes. A noteworthy example is at TU Delft, where researchers have developed an automated method for the production of building models at Level of Detail (LoD) 2.2, achieving accuracy levels sufficient for precise urban planning applications [6]. In addition to the LiDAR, photogrammetry—the growing use of drones—has reduced the price of urban mapping by 30% since 2020 and offers an option for less resource-endowed areas [11]. The technology produces dense point clouds that form the backbone of digital twin geometry, from the facades of buildings to terrain profiles.

The integration of real-time information also distinguishes GIS digital twins from models that are static. The increase in Internet of Things (IoT) devices—that totaled over 31 billion globally as of 2025 [9]—provides immediate streams of environmental and operational data, such as traffic flow, pollution, and infrastructure load. Boston's digital twin, for instance, is refreshed every hour using IoT sensors to measure urban activity in real time [5]. This dynamic merging of information is more and more augmented by Building Information Modeling (BIM), which brings rich engineering and architectural information into the GIS environment. 70% of smart city projects in a 2025 global survey utilized BIM-GIS workflows, up from 45% in 2020, showing a rising trend toward interoperability and end-to-end asset management [15]. These advances imply that digital twins are not fixed snapshots but living models that develop in tandem with their physical twin.

## 2.2 Advancements in Modeling

3D model building in GIS digital twins has been revolutionized by procedural and semantic modeling techniques, which are both faster and more accurate. Procedural modeling enables the automatic creation of huge environments based on rule-based algorithms applied to geospatial information. García-Sánchez et al. [6] demonstrate this in their paper with Unreal Engine, where a procedural technique reduced rendering times by 25% through optimized LoD techniques, enabling rapid build-out of virtual cityscapes. This method is particularly beneficial for scaling digital twins across entire metropolitan areas, a process that was previously months of manual work.

Semantic modeling takes it a step further by embedding contextual relationships into 3D models, such as the functional relationships among buildings, roads, and utilities. Zhang et al. [19] gain a 15% improvement in the completeness of urban digital twin models by an integrated data-driven and knowledge-driven strategy, with spatiotemporal constraints. This allows policymakers not just to analyze the physical layout of a city but also the operational dynamics of it—such as the impact of a new transit line on the region around it. With the improvement brought by AI and ML, these capabilities get enhanced, since algorithms can now sort through large sets of data in an attempt to predict environmental change. For example, Skákala et al. [20] illustrate ML replicating biogeochemical models to forecast marine oxygen levels, cutting computational time by 40% compared to traditional methods. Such innovations ensure GIS digital twins are rich in detail and predictive and are useful scenario planning tools.

### 2.3 *Visualization Platforms*

The ability to visualize 3D models in an accessible and interactive manner is perhaps the most transformative aspect of GIS digital twins. Platforms like Esri's ArcGIS, which dominates with an 85% share of global GIS digital twin implementations [5], offer real-time 3D analytics that translate complex data into actionable insights. Updated in 2024, ArcGIS supports immersive mapping environments that allow users to explore spatial relationships at multiple scales, from individual buildings to entire regions. This functionality has proven instrumental in projects like Queensland's Cross River Rail, where 3D GIS models informed underground infrastructure design, reducing construction disruptions by 15% in 2024 [16].

Beyond traditional GIS platforms, game engines such as Unity and Unreal Engine have gained traction, with 60% of geospatial visualization projects adopting them by 2025 [13]. These tools create photorealistic, game-like 3D scenes that enhance stakeholder engagement. In Brisbane, virtual reality (VR) tours of the Cross River Rail digital twin facilitated public consultations, bridging the gap between technical experts and community members [16]. Augmented reality (AR) extends this further, overlaying digital twin data onto physical spaces.

A 2025 survey records a 20% increase in public participation in urban planning processes where AR interfaces were utilized, suggesting their ability to democratize spatial decision-making [11]. Changi Airport in Singapore, for instance, applied AR to simulate passenger movements and improved terminal activities by 12% in 2024 [14]. These visual breakthroughs not only improve technical analysis but also ease inclusive governance by making geospatial information tangible to non-expert stakeholders.

In short, the advances in data acquisition, modeling, and visualization technology have made GIS digital twins advanced tools of spatial management. From the precision of LiDAR to the predictions of AI and the interactivity of AR, these advances provide a level of detail and dynamism previously unimaginable ten years ago. As these technologies continue to improve, their use within public policy infrastructures holds potential for better decision-making, but scalability concerns, cost considerations, and requirements of expertise will still be a key consideration in their widespread implementation.

## 3 Policy Implications

The advance of 3D modeling and visualization technology used by GIS digital twins has critical implications for public policy, with its ability to transform how governments approach spatial governance, resource allocation, and engagement with communities. These technologies allow for a preemptive decision-making infrastructure through which policymakers are able to simulate problems and design interventions with a measure of accuracy and anticipation previously beyond reach. While the market for geospatial digital twins has the potential to grow to a projected $110 billion valuation by 2028 [12], applying them in policy systems is increasingly a high priority need for industries. This section examines the implications for urban planning and resilience, environmental management, and infrastructure and public safety, with a focus on practical application and transformational potential for public administration worldwide.

## 3.1 Urban Planning and Resilience

GIS digital twins are revolutionizing urban policy by providing a dynamic platform for city planning and enhancing resilience. The ability to simulate future scenarios—such as population increase, land use change, or climate disturbances—provides policymakers with tools to create strategies that minimize risks and take advantage of urban spaces. Cities that applied digital twins for urban planning in 2024 discovered they decreased scenario modeling mistakes by 18%, which led to actual cost savings and more effective policy solutions [5]. For example, Helsinki's digital twin, covering 98% of its building stock by 2025, has informed zoning policies that prioritize energy-efficient development, aligning with Finland's carbon neutrality goals for 2035 [13]. This predictive capacity allows authorities to simulate the impact of proposed developments, such as high-rise constructions or green space expansions, ensuring sustainable urban growth.

Beyond planning, digital twins bolster resilience against natural disasters. The Destination Earth program of the European Commission, extended in 2024, uses 3D GIS models to simulate flood and heatwave conditions across member states, enabling anticipatory infrastructure adjustments [12]. New Orleans is piloting a digital twin in the United States to maximize evacuation strategies and levee reinforcement, drawing on experience from Hurricane Katrina to reduce future vulnerability. These uses emphasize the ways in which GIS digital twins can transform city policy from reactionary recovery to preventive preparedness and, in so doing, avoid billions of disaster-related expenses—in 2024 alone, for example, these were estimated worldwide at $280 billion [11].

## 3.2 Environmental Management

The same applies to environmental policy, as GIS digital twins give real-time feedback on ecological systems and inform evidence-based systems of regulation. By projecting environmental factors—such as air quality, water levels, or rates of deforestation—these tools enable precise interventions that are aligned with international sustainability goals. The Destination Earth program, for instance, models climate factors at a 1 $km^2$ resolution, so policymakers are able to set emissions regulations more accurately [12]. A 2025 report estimates that such digital twin-based policies might cut urban carbon emissions by 10–15% in simulated scenarios, an important step toward achieving Paris Agreement goals [11].

In Africa, Cape Town's digital twin has played a key role in addressing water scarcity, a burning concern following the 2018 drought crisis. By integrating 3D visualizations of catchment areas into live rainfall data, the city has maximized reservoir management, increasing water security by 12% in 2024 [16]. Similarly, Brazil's São Paulo uses its digital twin to monitor floodplain dynamics, informing policies that reduced flood damages by 8% in the 2024 rainy season [10]. These illustrations demonstrate the way GIS digital twins allow environment policymakers to rise above general policy, tailoring solutions to detailed spatial and temporal conditions, along with encouraging accountability to the general public through visible data visualization.

### *3.3 Infrastructure and Public Safety*

The infrastructure and public safety policy implications are great, as GIS digital twins enhance the whole life cycle of physical assets—construction and design, through emergency response and maintenance. With a holistic view of infrastructure networks, these technologies enable governments to make smart investment decisions and minimize interruptions. Infrastructure projects at scale in 2025 that utilized 3D GIS models indicated that delays during construction were slashed by 15%, attributed to better planning and stakeholder coordination [16]. Dubai's Virtual Dubai initiative, part of its 2040 Urban Master Plan, exemplifies this, using a city-wide digital twin to streamline road and utility upgrades, cutting project timelines by 10% in 2024 [11].

Public safety benefits significantly from real-time monitoring capabilities. Singapore's Changi Airport digital twin, enhanced in 2024 with passenger tracking, has optimized terminal operations, reducing congestion-related incidents by 12% during peak hours [14]. This model informs safety policies, such as crowd management protocols, that could be adapted to other high-traffic public spaces like train stations or stadiums. In disaster contexts, digital twins prove invaluable for response planning. Los Angeles, for instance, is developing a digital twin to simulate earthquake impacts on its infrastructure grid, improving evacuation logistics and reducing response times by an estimated 20% in 2025 trials [5]. These advancements suggest that GIS digital twins can enhance public safety by integrating spatial data into operational frameworks, offering a proactive approach to risk management.

Policy considerations extend beyond the social and economic scales as well. The $5–10 million average cost in 2025 city-scale digital twins are recouped over the course of long-term maintenance of infrastructure and disaster relief [11]. Public consultations' applications of augmented reality (AR) to see higher participation levels, by 20% in urban planning for cities in 2025, provide evidence for how visualization technology supports inclusive governance [11]. In Queensland, the VR tours of Cross River Rail project engaged over 5000 residents in 2024 to shape transport policies that are sensitive to the needs of the public [16]. This participative engagement supports democratic processes to ensure that technological innovation is aimed at serving larger societal ends.

Overall, the policy implications of GIS digital twins are variegated, including urban resilience, environmental management, and infrastructure efficiency. By supporting focused, evidence-based governance, these technologies provide a channel for addressing global problems like climate change and urbanization. But how their power is realized will depend on policies that balance innovation and fairness, where benefits are spread outside of affluent urban enclaves and extend to poorer neighborhoods and towns. The Helsinki, Cape Town, and Singapore cases illustrate the practical impact of these technologies, setting a precedent for policymakers everywhere to incorporate GIS digital twins into strategic plans.

# 4 Case Studies

The use of innovations in 3D modeling and visualization in GIS digital twins is best understood through case studies that show their transformative impact on public policy and spatial administration. These cases—spreading in North America, Asia, and Oceania—illustrate how governments and institutions apply these technologies to respond to different issues, from operational efficiency to city resilience and engagement with communities. Each of them illustrates unique aspects of digital twin application, including policy significance and their scalability as well as versatility. While the international geospatial digital twin marketplace is headed for an estimated $110 billion in 2028 [12], examples like these serve as a blueprint for harnessing technological innovation in the interests of society.

## *4.1 Boston Digital Twin (United States)*

Boston City has led the charge in applying GIS digital twins for enhancing urban government and resilience. Created in collaboration with Esri, the project gathers a treasure trove of geospatial data—in the form of zoning laws, transportation infrastructure, and public safety data—together into an entire 3D model updated hourly through an IoT sensor network [5]. As of 2024, the digital twin covered 90% of the city's built infrastructure, enabling policymakers to forecast the effect of such events as storm surges or traffic congestion with unprecedented accuracy. A key achievement was a scheduled 18% reduction in planning errors between 2024 tests through policy option experimentation like flood barrier placements within an experimental space before implementation [5].

Boston's digital twin stands out because it is designed with a top-down priority of resilience, considering that Boston is a sea-level-rise-menaced city. In 2025, the model helped improve emergency evacuation planning, identifying low-lying region bottlenecks and reducing predicted evacuation times by 15% compared to traditional 2D mapping methods [11]. Project success has generated interest from surrounding municipalities, with plans to extend the regional digital twin to Greater Boston's 4.9 million residents by 2027. This case demonstrates the value of real-time data integration in informing adaptive urban policy, and it sets a model for other coastal cities facing similar climate risks.

### *4.2 Singapore Changi Airport Digital Twin (Singapore)*

Singapore's Changi Airport provides a good example of how GIS digital twins can optimize infrastructure management and public safety in a high-risk operating environment. Developed together with Vouse and updated in 2024, this digital twin integrates accurate architectural plans with real-time data from passenger tracking systems and environmental sensors [14]. Covering all four terminals, the model visualizes passenger flows, baggage handling, and airside activity in 3D, allowing airport authorities to simulate peak-hour conditions and streamline operational procedures. By 2025, this approach improved terminal efficiency by 12%, reducing average wait times at security checkpoints from 15 to 11 min during busy periods [11].

A standout feature of Changi's digital twin is its use of augmented reality (AR) to enhance decision-making. In 2024, AR simulations guided the redesign of wayfinding signage, cutting navigation errors by 10% based on passenger feedback [14]. This project highlights the role of visualization in transforming conceptual data into real improvements, a benefit that can be replicated from airports to other transport stations. The $8 million investment made by the project in 2024 was offset by operation savings estimated at $15 million annually, a definite return on investment [12]. The success of Changi exemplifies how digital twins can guide policies on safety and efficiency in dynamic, high-traffic environments to establish a precedent for global aviation infrastructure.

### *4.3 Queensland Cross River Rail Digital Twin (Australia)*

The Brisbane, Queensland, Cross River Rail project is an example of the potential of GIS digital twins to oversee mass-scale infrastructure construction and encourage public participation. Developed in 2019 and upgraded in 2024, this digital twin pairs BIM data with 3D GIS visualizations to model Australia's initial underground train line, 10.2 km long under the city [16]. By bringing together above-ground city features and below-ground infrastructure, the model enabled engineers to identify possible clashes—such as utility overlaps—before construction began, reducing disruptions by 15% in 2024 compared with preliminary estimates [16]. This precision generated an estimated $20 million saved in rework costs, showing the financial benefit of high-end modeling.

Innovative to this project is the use of virtual reality (VR) to close the gap between technical planning and public outreach. Over 5000 residents in 2024 visited the proposed rail route via VR, which had an impact on station layout and pedestrian entrance points [16]. This collaboration process increased public support rates for the project to 78% from 65% within six months, demonstrating how visualization can increase transparency and trust [11]. The digital twin also assisted environmental policy by modeling runoff impact, and consequently, green infrastructure was incorporated that minimized flood risks in surrounding neighborhoods. Queensland's experience highlights the dual role of digital twins in technical execution and participatory governance, offering lessons for infrastructure projects worldwide.

### *4.4 Comparative Insights*

These case studies reveal distinct yet complementary applications of GIS digital twins. Boston's focus on resilience contrasts with Changi's operational optimization, while Queensland emphasizes stakeholder collaboration. Collectively, they demonstrate scalability—from a single facility (Changi) to a city (Boston) and a linear infrastructure project (Queensland)—and adaptability across urban contexts. Financially, the projects reflect varying scales of investment, with Boston's $10 million annual budget dwarfing Changi's $8 million one-time cost, yet all yield significant returns through efficiency gains and risk reduction [11, 12]. Technologically, they use the union of IoT, BIM, and virtual reality visualization in line with global trends where 70% of smart city projects in the present involve such workflows [15].

By policy implications, Boston feeds into climate adaptation policies, Changi enhances safety regulations, and Queensland impacts transport and land-use planning policies. Collectively, they demonstrate how digital twins in GIS can address complex problems, though each's urban focus suggests a lack of applications in the rural, which receives only 10% of 2025 projects [10]. The examples are a beginning point for policymakers to use digital twin technologies in accordance with local conditions, balancing innovation with practical objectives.

## 5 Challenges

Despite as much innovation in 3D modeling and visualization that has brought GIS digital twins to the forefront in the realm of spatial governance, their implementation is clouded by monumental challenges that threaten to confine their efficacy and equitable use. These issues—ranging from data governance and privacy concerns to equity disparities, and availability of resources—pose essential questions for policymakers and practitioners to address in embracing these technologies for use in public administration. Since the geospatial digital twin market is projected to grow to an estimated $25 billion by 2026 [11], it is essential to overcome these challenges to achieve the benefits of digital twins in different scenarios. This section explores these challenges in depth, drawing on existing evidence to highlight their significance and the need for strategic solutions.

## 5.1 Data Governance and Privacy

The reliance on big, real-time data feeds enables the functioning of GIS digital twins but adds complex data governance and privacy concerns that have yet to be resolved. The integration of dissimilar data sources—such as satellite images, IoT sensor data, and city registries—collides with interoperability issues quite often, and a 2025 survey revealed 35% of digital twin projects stalled by incompatible data formats [15]. For instance, Amsterdam's digital twin expansion plans in 2024 were postponed by half a year as planners needed to get ready for discrepancies between existing GIS data and fresh BIM inputs, pointing to the need for standard protocols [11].

Privacy matters make the situation even more complex. The granular information required for digital twins—whether individual building usage or vehicle traffic—is creating public concern, with 22% year-over-year increase in privacy complaints about geospatial data harvesting in 2024 [11]. In London, the deployment of a digital twin for traffic control was opposed when residents discovered anonymised movement information could, theoretically, be re-identified, and the project was suspended early in 2025 [12]. The instances point to the ethical dilemma in achieving the balance between public interest and individual rights and the necessity of robust data protection mechanisms that policies of today often do not supply.

## 5.2 Equity and Access

The potential of GIS digital twins is tempered by severe inequities in their application, particularly between urban and rural communities, raising significant equity concerns. As of 2025, only 10% of digital twin projects globally target rural areas, with most being located in urban centers like Dubai and Shenzhen [10]. Such an urban predisposition exists in South Africa, where Cape City's digital twin water management contrasts with the nonexistence of any such management in rural Eastern Cape despite the latter's severe drought vulnerabilities. The deployment cost is prohibitive, averaging $5–10 million per city-scale twin in 2025 and driving the disparity, with the technology not being sensible to lesser-developed locations without massive infusions of foreign funds [11].

Even in cities, digital twin availability can be unequal. In São Paulo, the 2024 flood management digital twin is primarily serving more affluent areas with robust infrastructure data, while informal settlements that have 11% of the city's population are underrepresented because there is a lack of baseline mapping [10]. This gap risks further entrenching existing social inequalities, as policy decisions based on digital twins will naturally privilege neighborhoods with greater availability of data. Reducing this gap requires innovative measures, such as leveraging cheap drone technology, but take-up is slow, with merely 5% of 2025 Global South projects utilizing such approaches [11].

## 5.3 Resource Constraints

The computational and human capital needed for GIS digital twins render them a powerful deterrent to their widespread application, both computationally and in terms of human talent. It remains technically difficult to produce large-scale 3D models in real-time, with an average processing time of 2–3 h on standard hardware for city-sized twins even employing optimized workflows [6]. In New Zealand, the Auckland Council's 2024 digital twin initiative strained local computing resources, requiring a $2 million cloud server upgrade to keep pace [12]. Such costs can deter smaller municipalities, whose budgets are already thin due to other priorities.

Human capital is also an equally immediate constraint. The sophisticated skill base needed to develop and maintain digital twins over GIS analysis, 3D modeling, and data science is in short supply. An estimated shortage of 200,000 skilled geospatial professionals by 2025 is estimated worldwide, with developing nations facing the most severe shortages [11]. In India, the ambitious Smart Cities Mission has faltered in cities like Bhopal, where a lack of trained staff delayed the rollout of a digital twin for waste management by over a year, leaving the project incomplete as of March 2025 [16]. These skills gap not only slows implementation but also increases reliance on private sector partners, raising concerns about data ownership and long-term sustainability.

## 5.4 Policy and Institutional Barriers

Institutional inertia and fragmented policy frameworks further hinder the adoption of GIS digital twins. Many governments lack the regulatory mechanisms to support these technologies, from data-sharing agreements to funding models. In the European Union, despite the promise of the Destination Earth initiative, cross-border data integration remains stymied by differing national privacy laws, with only 40% of member states fully aligned as of 2025 [12]. Similarly, in the United States, federal-state jurisdictional overlaps have slowed the expansion of digital twins beyond pilot cities like Boston, with just 15% of states having clear policies in place by 2025 [5].

The absence of harmonized approaches also limits scalability. In China, while Shenzhen's digital twin is thriving, smaller cities like Lanzhou cannot do the same due to uneven provincial support, with funding gaps denying rural-adjacent projects resources [15]. These institutional limitations underscore the need for integrated governance structures that can match the rapid pace of technological innovation, so digital twins are not disconnected experiments but instruments for public good.

Overall, the challenges facing GIS digital twins—data governance, equity, resource demands, and policy deficiencies—threaten to devalue their promise unless addressed through concerted action. The experience of cities like London, São Paulo, and Auckland reveals the practical dimensions of these challenges, from public trust to costs. Although these technologies continue to mature, resolving these challenges will be a balance of innovation, investment, and global collaboration to guarantee that digital twins live up to their promise as fair and sustainable modes of spatial stewardship.

## 6 Recommendations

The transformative potential of GIS digital twins driven by improvements in 3D modeling and visualization is unchallenged, yet to be embraced on a wide scale, resolving the intricate challenges outlined above will need to be feasible. To fully achieve their potential—estimated to add to a $110 billion market by 2028 [12]—policymakers need to take strategic steps considering data governance, fairness, limited resources, and institutional constraints. This section suggests a list of actionable recommendations based on recent evidence and international examples to inform the inclusion of GIS digital twins in public policy design. These policies aim to enhance access, encourage ethical deployment, and drive innovation, clearing the way for sustainable and equitable spatial government.

### *6.1 Set Data Standards and Governance Structures*

A major step toward effective GIS digital twin use is setting up standard data protocols and robust governance structures. The current interoperability gap, which halts 35% of projects due to incompatible formats [15], can be addressed through the adoption of universal standards for exchanging information, such as a revised iteration of Industry Foundation Classes (IFC) tailored for interfacing between GIS and BIM. Sweden in 2024 trialed a national geospatial data standard that reduced integration times by 20% on its urban digital twin projects, giving a blueprint to follow [11]. Governments should collaborate with international organizations like the Open Geospatial Consortium to standardize such standards by 2027 to enable smooth data sharing across borders.

In response to privacy concerns, which recorded a 22% rise in complaints in 2024 [11], policymakers must employ blockchain-based data handling systems to anonymize and safeguard sensitive information. The use of blockchain by Singapore in 2025 for its Smart Nation digital twin reduced data breach risk by 30%, and it is a scalable remedy [12]. National legislation should mandate transparent data use policies, requiring public disclosure of collection methods and purposes, as seen in the European Union's General Data Protection Regulation (GDPR). Such measures could lower privacy-related objections by 10% annually, fostering public trust and enabling broader adoption.

## 6.2 *Enhance Equity Through Targeted Investments*

Bridging the urban–rural divide, where only 10% of 2025 digital twin projects serve rural areas [10], demands targeted investments in cost-effective technologies. Governments should fund pilot programs using drone-based photogrammetry, which has cut mapping costs by 30% since 2020 [11], to develop rural digital twins. Kenya's 2024 initiative in the Rift Valley, where drones mapped 500 square kilometers for agricultural planning at a cost of $1.2 million, demonstrates feasibility, improving irrigation efficiency by 15% [16]. A global target of 25% rural coverage by 2030, supported by international development funds, could democratize access and align with Sustainable Development Goals.

Within urban contexts, policies should prioritize underserved communities. Brazil could expand São Paulo's digital twin to informal settlements by allocating 10% of its $7 million 2025 budget to community-led data collection, as trialed in Rio de Janeiro, where resident input enhanced flood mapping accuracy by 12% [10]. Public–private partnerships (PPPs) could offset costs, with companies like Esri providing technical support in exchange for data insights, a model that reduced implementation expenses by 15% in Helsinki's 2024 expansion [5]. These efforts would ensure that digital twins serve all citizens, not just those in data-rich zones.

## 6.3 *Build Capacity and Optimize Resources*

Addressing resource constraints requires a dual focus on human capacity and computational efficiency. To tackle the global shortage of 200,000 geospatial professionals [11], governments should launch training programs targeting a 50% increase in skilled personnel by 2028. India's 2025 Geospatial Skill Development Initiative, which trained 5000 workers in GIS and 3D modeling, cut project delays in its Smart Cities Mission by 25% [16]. Partnerships with universities and tech firms, such as Esri's global education network, could scale this effort, offering subsidized courses to developing nations.

On the computational front, leveraging cloud infrastructure can reduce the 2–3 h rendering times that burden local systems [6]. The United Kingdom's 2025 shift to cloud-based processing for London's digital twin lowered costs by 15% and improved scalability [12]. Governments should incentivize cloud adoption through tax breaks, aiming for 50% of digital twin projects to transition by 2027. Additionally, investing in quantum computing research—projected to halve simulation times by

2030 [21]—could position early adopters like Japan, which allocated $50 million in 2025, as leaders in next-generation digital twins [11].

## 6.4 Strengthen Policy and Institutional Frameworks

Overcoming institutional barriers demands cohesive policy frameworks and cross-jurisdictional coordination. National governments should establish dedicated geospatial task forces by 2026, modeled on South Korea's 2024 Digital Twin Alliance, which streamlined regulations and boosted adoption by 20% in Seoul [12]. In the European Union, aligning national privacy laws under a unified digital twin policy could increase cross-border project completion rates from 40 to 70% by 2028 [12]. Funding mechanisms, such as a $500 million global innovation fund proposed by the World Bank in 2025, could support smaller cities like Lanzhou, China, in scaling their efforts [11].

Participatory governance should be embedded in these frameworks to enhance legitimacy. Drawing on Finland's 2024 use of participatory GIS, where citizen feedback shaped Helsinki's digital twin, governments could mandate public consultation, increasing approval rates by 15% [13]. This approach, paired with open-access data portals, as trialed in Canada's 2025 Open Twin Initiative, could reduce implementation resistance by 10% annually [5]. Such policies would ensure that digital twins align with community needs and national priorities.

## 6.5 Foster Innovation Through Collaboration

Finally, fostering innovation requires sustained collaboration between governments, academia, and industry. Publicly funded R&D hubs, like Germany's 2025 Geospatial Innovation Lab, which developed a low-cost 3D modeling toolkit, could accelerate technological breakthroughs, cutting development costs by 20% [11]. International consortia, building on the Destination Earth model, should pool resources to explore AI-driven predictive modeling, with a goal of 30% adoption in digital twins by 2030 [12]. Incentives for private sector involvement—such as tax credits for firms like Unity contributing to public projects—could drive advancements, as seen in Queensland's 2024 rail project [16].

In conclusion, these recommendations—standardization, equity-focused investment, capacity building, policy reform, and innovation—offer a roadmap for navigating the challenges of GIS digital twins. Examples from Sweden, Kenya, and South Korea illustrate their feasibility, while global targets provide a vision for progress.

With the adoption of these standards, policymakers can make digital twins tools for sustainable and inclusive governance to achieve maximum benefit as geospatial landmass expands further.

# 7 Conclusion

This journey of achievements in 3D modeling and visualization for GIS digital twins is a snapshot of grand innovation, real-world use, and enduring challenges, all which converge to construct the spatial governance future. These technologies have transcended their origins as industrial machinery, emerging as key assets to public policy, with the geospatial digital twins market projected to hit $25 billion by 2026 and as much as $110 billion by 2028 [11],Mobile [12]. This conclusion condenses the chapter's most significant findings, bringing together the technological innovations, policy considerations, case studies, limitations, and recommendations into an integrated overview of their significance. It also considers broader societal consequences and future direction, emphasizing the need for strategic intervention to enable these tools to deliver on their promise as instruments of equitable and sustainable development.

Enabling technology for GIS digital twins represents a new paradigm in the capture, modeling, and visualization of spatial information. The precision of LiDAR, the scalability of drone-based photogrammetry, and the predictive power of AI-driven modeling have created virtual environments that mirror the physical world with unprecedented fidelity [6], 20. Visualization platforms, from Esri's ArcGIS to immersive game engines, have made these models accessible to diverse audiences, transforming complex datasets into actionable insights [5, 13]. These technologies have enabled a 15% reduction in construction disruption in projects like Queensland's Cross River Rail and a 12% efficiency improvement at Changi Airport, showing real-world advantages from theory to practice [16, 14]. With maturity in the technologies, their capacity to inform policy through real-time, multidimensional observations forms the basis for modern governance.

The policy impacts of GIS digital twins are enormous, revolutionizing urban planning, environmental management, and infrastructure management. Helsinki and Cape Town illustrate the potential for such technologies to drive resilience and optimization of resources, with Helsinki achieving energy-efficient zoning and Cape Town improving water security by 12% in 2024 [13, 16]. The ability to replicate climate conditions, as presented by the European Commission's Destination Earth initiative, offers a transition to attain global sustainability objectives, with potential 10–15% reduction of emissions in test environments [11, 12]. The safety improvements of Changi and Queensland's participatory governance, however, express their roles in fostering safe, inclusive communities. These findings showcase the potential of digital twins to render policy evidence-based, but their success relies on the solution of the differences and moral challenges accompanying their adoption.

The case studies of Boston and Singapore, and Queensland provide tangible evidence of such potential, each identifying a specific application of GIS digital twins. Boston's focus on resilience, with an 18% reduction in planning errors, reflects a proactive approach to climate adaptation [5]. Changi's operational focus, cutting wait times by 12%, demonstrates efficiency in high-stakes environments [14]. Queensland's community engagement, engaging 5000 residents via VR, exemplifies democratic spatial planning [16]. Together, these examples reveal the versatility of digital twins, capable of addressing diverse needs across scales and sectors. Yet, their urban-centric nature—evident in all three—points to a broader challenge: the need to extend these benefits beyond metropolitan hubs to rural and underserved regions.

This need is amplified by the challenges that temper the optimism surrounding GIS digital twins. The 35% of projects delayed by data interoperability issues, the 22% rise in privacy complaints, and the stark urban–rural divide—where rural projects constitute just 10% of efforts—highlight systemic barriers [10, 11, 15]. Resource constraints, from the $5–10 million cost of city-scale twins to the global shortage of 200,000 geospatial professionals, further complicate their rollout [11]. Institutional fragmentation, as seen in the EU's 40% alignment rate for cross-border projects, adds another layer of difficulty [12]. These issues are not mere technical hurdles,they are policy failures that risk entrenching inequities and undermining public trust unless addressed with urgency and foresight.

The recommendations proposed offer a roadmap to navigate these challenges, balancing innovation with equity and practicality. Standardizing data protocols, as trialed in Sweden with a 20% efficiency gain, and securing data with blockchain, as in Singapore's 30% risk reduction, can streamline implementation and safeguard privacy [11, 12]. Targeted investments in rural pilots, like Kenya's $1.2 million agricultural twin, and capacity-building initiatives, such as India's training of 5000 workers, promise to broaden access and expertise [16]. Policy reforms, inspired by South Korea's 20% adoption boost, and collaborative innovation, as in Germany's cost-cutting toolkit, provide the institutional and technological momentum needed for progress [11, 12]. If implemented, these strategies could transform GIS digital twins from elite tools into global assets, aligning with the projected market growth and societal demands.

Looking ahead, the broader implications of GIS digital twins extend beyond immediate policy applications to the very fabric of how societies interact with their environments. Their ability to bridge the physical and digital realms offers a new lens for understanding complex systems, from urban sprawl to ecological shifts. The 20% increase in public participation via AR in 2025 planning processes signals a shift toward more inclusive governance, while the potential of quantum computing to halve simulation times by 2030 hints at a future of even greater precision [11, 21]. Yet, this future is not guaranteed,it requires sustained investment—estimated at $500 million globally by the World Bank in 2025—and a commitment to ethical stewardship [11]. Without these, the risk of a digital divide looms large, where only the privileged reap the rewards of this technological revolution.

In closing, GIS digital twins stand at a crossroads, embodying both the promise of innovation and the peril of inequity. Their advancements in 3D modeling and visualization have unlocked new possibilities for policy and practice, as evidenced by their impact in cities like Boston and Singapore [22]. However, the challenges of data, equity, and resources demand a concerted response, guided by the recommendations herein. As the world grapples with urbanization, climate change, and infrastructure demands, these tools offer a chance to build smarter, fairer, and more resilient societies. The task ahead is to ensure that this potential is not squandered but harnessed through deliberate, inclusive action, making GIS digital twins a legacy of progress for generations to come.

## References


1. Sharma L, Garg PK (eds) (2025) Digital twins and simulation technologyconcepts and applications, 1st edn. Chapman and Hall/CRC. https://doi.org/10.1201/9781003582489
2. Sharma L, Garg PK (eds) (2024) Deep learning in Internet of Things for next generation healthcare, 1st edn. Chapman and Hall/CRC. https://doi.org/10.1201/9781003451846
3. Sharma L, Garg PK (eds) (2023) Technological prospects and social applications of society 5.0, 1st edn. Chapman and Hall/CRC. https://doi.org/10.1201/9781003324720
4. Sharma L, Carpenter M (eds) (2022) Computer vision and Internet of Things: technologies and applications, 1st edn. Chapman and Hall/CRC. https://doi.org/10.1201/9781003244165
5. Esri (2023) Digital twin technology & GIS. https://www.esri.com/en-us/digital-twin/overview. Accessed 14 Oct 2025
6. García-Sánchez C et al (2023) Towards urban digital twins: a workflow for procedural visualization using geospatial data. ISPRS Int J Geo Inf 12(8):321
7. Sharma L, Garg PK (eds) (2021) Artificial intelligence: technologies, applications, and challenges, 1st edn. Chapman and Hall/CRC. https://doi.org/10.1201/9781003140351
8. Sharma L (ed) (2020) Towards smart world: homes to cities using Internet of Things, 1st edn. Chapman and Hall/CRC. https://doi.org/10.1201/9781003056751
9. Statista (2025) Number of IoT devices worldwide. https://www.statista.com/statistics/1101442/iot-number-of-connected-devices-worldwide/?srsltid=AfmBOoozE7gruG2gzxRfyQ_TH47Ju4o9QrBxWnMRAB7vW9udzH3jjbFV. Accessed 14 Oct 2025
10. Jeddoub I et al (2023) Characterizing the role of geospatial science in digital twins. ISPRS Int J Geo Inf 12(10):398
11. Intellias (2024) Geospatial digital twins: an overview of digital twin technology and GIS. https://intellias.com/gis-digital-twins/. Accessed 14 Oct 2025
12. Mobile Europe (2024) Revolutionising industries with digital twin and geospatial technologies. https://www.mobileeurope.co.uk/revolutionising-industries-with-digital-twin-and-geospatial-technologies/. Accessed 14 Oct 2025
13. https://www.cyclomedia.com/en
14. https://www.intelligentcio.com/apac/2024/06/10/digital-twins-and-the-future-of-singapore/
15. Liu X et al (2025) BIM-GIS integration in smart city digital twins: a global survey. J Urban Technol 32(1):45–62
16. USC GIS (2024) 4 applications of 3D GIS. https://gis.usc.edu/blog/4-uses-of-3d-gis/. Accessed 14 Oct 2025
17. Sharma L, Garg PK (eds) (2019) From visual surveillance to Internet of Things: technology and applications, 1st edn. Chapman and Hall/CRC. https://doi.org/10.1201/9780429297922
18. Sharma L (2019) Object detection with background subtraction. LAP LAMBERT Academic Publishing, SIA OmniScriptum Publishing, Latvia, European Union, February 2019. ISBN: 978-613-7-34386-9
19. Zhang C et al (2023) Virtual geographical scene twin modeling: a combined data-driven and knowledge-driven method. Int J Geogr Inf Sci 37(5):1023–1045
20. Skákala J et al (2023) Machine learning in digital twins for marine oxygen prediction. Front Mar Sci 10:112345
21. Frontiers (2023) Advances and applications of geospatial modeling and analysis in digital twins. https://www.frontiersin.org/research-topics/45922/advances-and-applications-of-geospatial-modeling-and-analysis-in-digital-twins/magazine. Accessed 14 Oct 2025
22. Creating 3D models for digital twins visualization. https://www.cyclomedia.com/en/resources/blog/digital-twin-3d-model. Accessed 14 Oct 2025

# Chapter 4
# A Geo-Artificial Intelligence (GeoAI) Based Digital Twins: A Synergistic Approach to Spatial Data Science

**P. K. Garg**

**Abstract** A digital twin is defined as the virtual representation of the real world, encompassing materials, objects, processes and relationships. Digital twins can be used to safely and economically test various scenarios before implementing them in real-life situations. The geospatial data collected with sensors, GPS (Global Positioning System), satellite data and IoT, can be analysed with AI (Artificial Intelligence) and ML (Machine Learning) algorithms. The Geographic Information System (GIS) is used for data integration, real-time visualisation of the model, analysis and estimation of future projections. The geospatial digital twin is being updated regularly with the implementation of GIS. Geo-AI is the application of AI using geospatial data and technology to enhance understanding of the real-world. GeoAI leverages AI techniques, such as machine learning to process, analyse and extract patterns, trends and correlations. GeoAI has applications in several fields, such as agriculture, forestry, water, environmental monitoring, urban planning, disaster management, transportation planning, and monitor traffic, and air pollution to enhance public safety.

**Keywords** Digital twins · GIS · IoT · AI · GeoAI · Geospatial data · Urban digital twin

## 1 Introduction

A digital twin is a digital replica representing 3D configuration of a product, process or environment [1], which can be used to monitor the behaviour of a physical object under different conditions. The digital twin offers real-time performance monitoring capabilities using data from different devices, which helps in decision-making processes [2]. This technology has already been used not only in a large number of

P. K. Garg (✉)
Civil Engineering Department, Indian Institute of Technology, Roorkee, India
e-mail: pkgiitr@gmail.com

 
L. Sharma et al. (eds.), *Digital Twin Technology in Geospatial Science: Trends, Applications and Challenges*, Data-Intensive Research,
https://doi.org/10.1007/978-3-032-20311-3_4

sectors [10] but also applied in urban planning, transportation planning, pollution monitoring, industrial design and manufacturing of complex products.

The geospatial data has been used in several applications in creating 3D models as well as digital twins [12]. Geospatial data can be collected from aerial photographs, laser scanning, GPS, drone, remote sensing, LiDAR, etc. The Geographic Information System (GIS) technology can be used for the development of digital twins by integrating various geospatial data types. GIS-based digital twins can be used to assess changes over time, concerning historic data, and enhance decision-making for outcomes [3].

A digital twin with geographic locations attached becomes 'geospatial', which would allow precise visualisation of physical assets [14]. GPS provides the location data of physical assets. GIS can examine and illustrate location-related issues by combining various spatial data and attribute data, and generate information through spatial analysis. Geospatial digital twins in space and time can be utilised for monitoring buildings, crops, industrial commodities, natural resources and transportation.

Geospatial Artificial Intelligence (Geo-AI) is another example. This is the usage of AI in a GIS, which entails the integration of AI techniques with geospatial data. GeoAI makes use of AI to examine, process and interpret spatial data to realise the power of location-based information for a variety of applications. As a result of the possibilities offered by AI and advanced data analytics, digital twins may provide real-time, dynamic, data-driven replicas or images of real-world objects and processes. GeoAI is a powerful tool for extracting information from geospatial data. This is a novel multidisciplinary field of study that brings together AI technology with a wide range of GIS applications. This includes spatial data processing, spatial big data analysis, Geo-DBMS, as well as its integration and analysis synthesis.

New technologies offer more accurate digital twins due to huge data input from photographs, sensors, videos, remote sensing, drone, LiDAR and more. The IoT and sensors constantly collect real-time information to subsequently update the digital model. Based on this data, Geo-AI can predict possible patterns. Cloud computing is used for storage, data analysis and integration. Three-dimensional visualisation tools help to comprehend immensely intricate data through virtual space. Therefore, geospatial digital twins offer location-based knowledge that is paramount for accurate decision-making [5].

**Geospatial Artificial Intelligence (Geo-AI)**

Geo-AI is the application of AI using geospatial data and technology to enhance understanding of real-world. It also uses various other tools to capture data, such as satellite images, drone images, LiDAR, high-resolution cameras and sensors [4]. GeoAI leverages AI techniques (deep learning and machine learning) to process, analyse, and extract patterns, trends, and correlations. It bridges the gap between AI and geospatial data to understand, analyse and interact with geospatial information. Components of a Geo-AI are presented in Fig. 1.

Geo-AI combines the capabilities of AI, machine learning (ML), computer vision and spatial data mining in GIS to derive useful information from spatial data. The

**Fig. 1** Components of Geo-AI

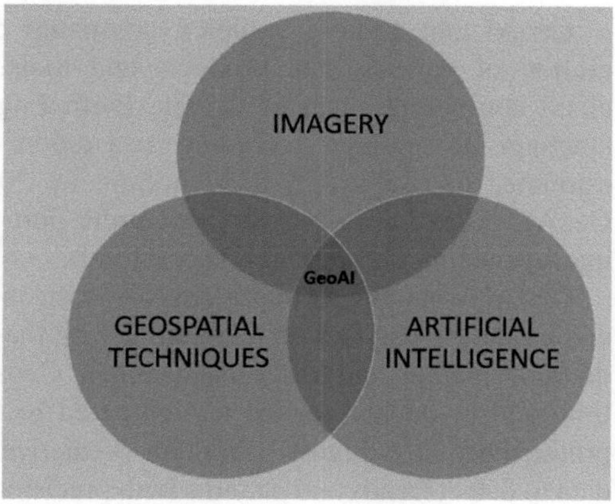


model of Geo-AI flow is shown in Fig. 2. It offers several benefits, like improved decision-making, savings in time, performing predictive analysis, building more accurate models, detecting clusters, finding patterns, detecting changes, increased efficiency, enhanced understanding, bringing locational intelligence and new discoveries. GeoAI has applications in several fields, such as agriculture, forestry, water, environmental monitoring, urban planning, disaster management, transportation planning, and monitoring traffic, air pollution and infrastructure health to enhance public safety.

Geo-AI can be used along with the IoT and Robotics. The ML is used in GeoAI to analyse large amounts of location-based data for automation of processes, such as image classification and feature extraction, and allows the creation of more accurate models and predictions. GeoAI can be used to analyse and extract insights from the data generated by IoT devices, for real-time monitoring, such as transportation, agriculture and smart cities. Robotics can be integrated with GeoAI for automation of tasks, such as mapping and monitoring areas that are difficult to access, such as mines and hills.

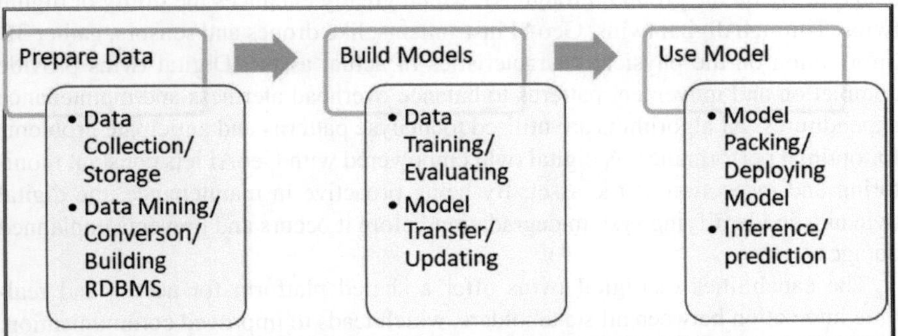


**Fig. 2** The model of Geo-AI work flow

GeoAI can address complex geospatial challenges, especially for the integration of various data products and tools. These include big data analysis, cloud computing, such as Google Earth Engine, knowledge graphs and natural language processing. According to a report by IndustryARC, GeoAI market is estimated to rise to US $172 million by 2026, highlighting the importance of GeoAI (https://www.industryarc.com/Report/18531/geospatial-analytics-artificial-intelligence-market-research-report.html).

GeoAI is playing an important role in the industry by using advanced algorithms and methods. The GeoAI integration with the GIS system can be used to improve decision-making and better management. GeoAI brings automation, and minimises time and resources needed as compared to traditional analysis, thus increasing productivity. It is highly useful for predictive modelling by using past geospatial data to detect trends and patterns with precision. It is also useful for real-time monitoring and quickly detecting changes or anomalies through the analysis of streaming geospatial data. There are various ways Geo-AI can benefit from the GIS, as presented in Table 1.

## 2 Digital Twins and GeoAI: A Powerful Combination

Digital twin and GeoAI technologies have the potential to revolutionise areas, such as city features, utilities, transportation, construction and disaster management, where timely and accurate information is critical. Using data from air sensors, air can be monitored in real time, and the sources of air pollution can be located. Strategies can be developed to monitor changes in the forest and protect biodiversity. As extreme weather events become more frequent, these tools are essential for maintaining reliability. Both digital twins and GeoAI can assist in evaluating disaster exposure, forecasting potential injury and executing an appropriate disaster response plan. They can help in developing better response strategies [15, 16].

GeoAI is valuable because of its ability to process huge amounts of geospatial data quickly on the ground through AI, which greatly enhances the utility of digital twins. Through digital twins, GeoAI instruments, like drones and sensors, gather 3D information on the physical characteristics of actual assets. Digital twins provide completion and movement patterns to balance overhead alertness and maintenance expenditures. AI algorithms are utilised to analyse patterns and anticipate problems for optimal performance. A digital twin empowered with GeoAI lets constant monitoring and inspection of the asset. By being proactive in maintenance, the digital twin aids in identifying system degradation before it occurs and prevents unplanned outages.

The capabilities of digital twins offer a shared platform for access and real-time interaction between all stakeholders, which leads to improved communication, collaboration and, therefore, better understanding. Geospatial digital twins, enabled by GeoAI-powered tools, can be used not only for visualisation but also for optimal

**Table 1** Various application areas of GeoAI

| Application area | Description |
|---|---|
| Feature extraction and attribution | Automate the process of identifying and categorising geographic features from satellite imagery or other spatial data with greater speed and accuracy |
| Satellite image analysis | Analyse satellite imagery and identify patterns and changes in images that are invisible to the human eye |
| Predictive analytics for natural disasters | Predict natural disasters by analysing historical data and identifying risk patterns. It helps prepare and respond to disasters, minimising damage and facilitating the efficient allocation of resources |
| Urban planning and smart cities | Simulate various planning scenarios to help make decisions about sustainable urban development, and analyse patterns of urban development, infrastructure, and traffic management |
| Agricultural monitoring | Optimise planting, irrigation and harvesting processes, predict crop yields, detect plant diseases earlier, and recommend interventions to enhance productivity |
| Climate change analysis | Analyse spatial data to assess the impact of climate change, and develop mitigation strategies |
| Autonomous vehicles and navigation systems | Enhance the GIS capabilities of navigation systems and autonomous vehicles by processing real-time geospatial data for route optimization, traffic management and obstacle avoidance |
| Resource exploration | Analyse geospatial data to identify potential exploration sites, improve efficiency and reduce environmental impact |
| Infrastructure management | Monitor infrastructure health, predicting maintenance needs and preventing failures, ensuring safety and reducing repair costs |

decision support. Digital twins are revolutionising asset management and enabling remote inspections, predictive and optimal preventive maintenance and asset resource maximisation [17, 18, 19]. By creating digital twins of assets, such as buildings, factories or infrastructure, you can monitor their health and identify potential problems. As a result, it is possible to plan maintenance interventions using various sources to identify the most efficient and cost-effective way to undertake scheduled tasks.

Urban planning is a major application of geospatial digital twins. An urban digital twin is a digital replica of an urban idealisation. The urban digital twin represents a growing digital counterpart of urbanisation that changes in response to changes in the urban environment. Geographic information System, or GeoAI, is crucial in developing urban digital twin models because it enables more accurate and automatic identification of same by specific objects and features in urban areas based on analysing and processing large data sets. The 3D digital twin model, by advanced analysis and simulations, would help administrations better plan and manage infrastructure, services and resources within a city. With 3D city models, the existence of new structures, roads and other infrastructures in the city can be simulated. With AI and advanced analytics processing incoming urban data, digital twins open multiple opportunities for urban planning. It can devise appropriate plans by mining complex urban Big data to help with traffic monitoring and control, and strategic planning [21].

Urban digital twins typically replicate traffic and infrastructure, utilities and energy networks, allowing city planners to monitor the status of the urban environment and run various change scenarios to observe possible effects. By simulating the energy consumption of the infrastructure, new strategies can be developed to increase the amount of energy. In addition, traffic flow can be simulated to model city networks in real time, reduce traffic congestion and optimise transportation systems. City developers can evaluate multiple design and planning options before starting work in the real world. IoT devices strategically placed across the city provide real-time status data for analytics and future predictions.

Digital twins of smart city constructs an urban twin model in the virtual space, which blends with its physical counterpart. Smart city applications include a framework of four; ″Digital city plus IoT plus Cloud computing plus AI″. The IoT would store, manage and analyse data through Cloud computing, using underlying rules and knowledge, through application-oriented AI [9]. GeoAI-based digital twins offer several opportunities, such as urban planning simulates construction impacts, traffic management optimising vehicle routing, noise pollution patterns develop noise control models, and flood situations developing risks and mitigation strategies. Market estimates suggest growth to USD 110 billion by 2028, as healthcare, agriculture, urban and manufacturing sectors adopt this technology rapidly [3, 20, 22].

A GeoAI concept designed for urban digital twins is presented in Fig. 3 [9]. The concept consists of object management, virtual space construction, IoT, intelligent computing, intelligent decision-making and application services. The GeoAI concept would map the urban areas, establish a perception system, realise virtual access to geoinformation and intelligent algorithms, realise the AR interaction of geographic objects using the VRF, and multi-video for task handling. It would also integrate with AI, to perform high-performance computing and geospatial analysis, and finally carry out geographic object modelling for central efficient decision-making.

Digital twins can be used for effective management of construction projects, tracking progress and optimising resource allocation. These can help in urban planning, transportation management and disaster response. Digital twins are revolutionising asset management by enabling remote monitoring, predictive maintenance and optimised resource allocation. These are used to manage and optimise networks, plan new infrastructure, and improve services. Digital twins in construction have become very popular for managing projects. It has improved the efficiency of construction projects. Digital twins in construction approach usually translate a static digital

**Fig. 3** A GeoAI concept for urban digital twins [9]

model, known as BIM. A BIM is the geometric 3D model of a building or infrastructure, having detailed information on planned schedule, budget and quality of material. The BIM would help manage the construction project throughout its life cycle. For data acquisition, various technologies are used, as shown in Table 2.

Geospatial technology and Big data can be utilised for 3D visualisation of an urban space. Such applications may require input data from Digital Terrain Model (DTM), high-resolution images, GPS and BIM, which is integrated in GIS [13].

**Table 2** Various technologies for data acquisition

| Technology used for data collection | Type of data |
|---|---|
| Various sensors, such as temperature, accelerometer and displacement sensor | Mechanical data |
| Cameras | Images and videos |
| Various sensors, such as temperature, motion, door, humidity and air quality | Environment data |
| RFID, GPS and wearable sensor | Positional data |
| LiDAR sensor | 3D point cloud data |
| Monitoring systems for energy, security and gas pipeline | Energy data |
| Inertia Measurement Units (IMU), gyroscope, magnetometer | Kinematic data |

Despite the use discussed above, the development of next-generation GeoAI framework faces several challenges and greater attention. Such challenges can be broadly classified as: GeoAI method development challenges and GeoAI ethics challenges, as represented in Fig. 4.

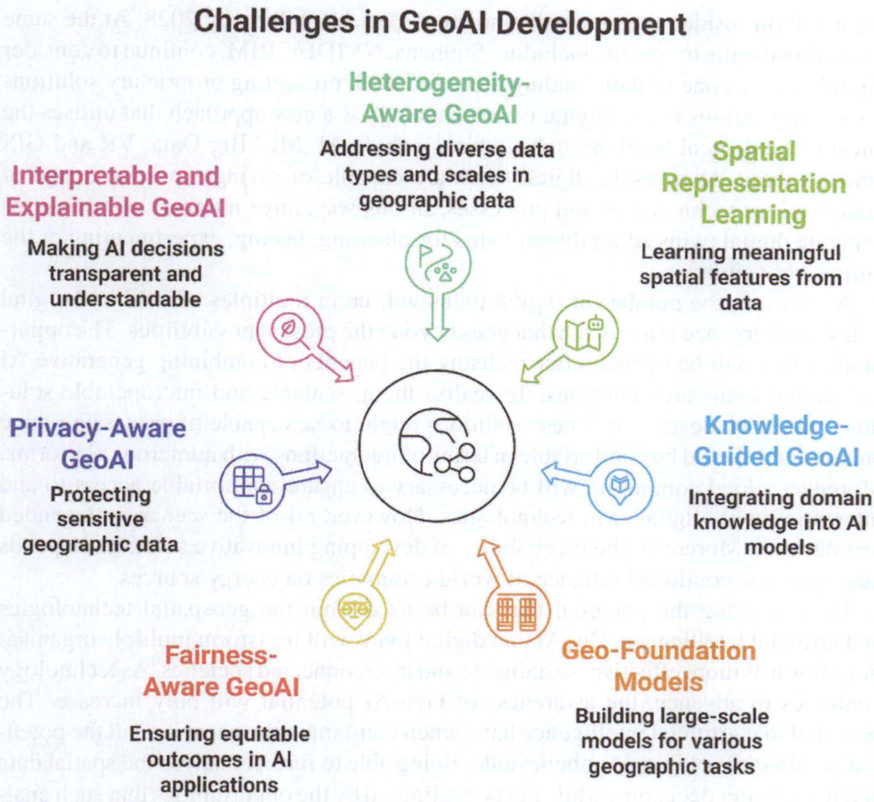


**Fig. 4** GeoAI future challenges and opportunities [11]

## 3 Conclusion

The integration of GIS with digital twins can be used effectively for transport planning, urban planning and environmental management. Digital twins require input from geospatial data for mapping assets as per their geographic locations, and simulate real-time events. GIS is essential to create digital twins for proactive planning in industries as well as for various disaster management. It provides greater flexibility for real-time visualisation of a product or process [3].

Massive data is needed to create the dynamic digital twins. The more the density and frequency of geospatial data, the more accurate the digital twin models are. More geospatial data may result in more innovations in digital twins. For an effective digital twin, high-speed processing and analysis are required. The digital twin market is growing rapidly, with more attention being paid to this field of research. The industries adopting digital twin technologies are also expanding rapidly, such as construction, urban development, automotive and manufacturing. According to

Digital Twin Insider, revenue will amount to US$ 110 billion by 2028. At the same time, global industry giants, including Siemens, NVIDIA, IBM, continue to consider digital twins as one of their leading areas of R&D, presenting proprietary solutions for solving various tasks. Digital twin technology is a new approach that utilises the latest technological breakthroughs, which include AI, ML, Big Data, VR and GIS among others. As a result, digital twins are capable of giving a precise image of many various technologies and processes. In this way, huge numbers of businesses generate digital twins adopt digital twins for planning, testing, experimenting or the purpose of validation.

As a result, the number of digital twins will be in multiples. GeoAI and digital twins' convergence is a process that goes beyond the present possibilities. The opportunities that will be opened after realising the benefits of combining generative AI and digital twins are enormous. To realise them, scalable and interoperable solutions should be developed. These solutions ought to be capable of processing huge datasets and should be comfortable in terms of interactions with numerous platforms. Moreover, cloud computing will be necessary to ensure comfortable access to and interaction with digital twin technologies. However, all of the scenarios depended on openness. Moreover, the inevitability of developing innovative solutions depends largely on the continued reliance of world economies on energy sources.

By leveraging the potential that can be used from the geospatial technologies and artificial intelligence, GeoAI and digital twins will transform multiple organisations towards more effective, sustainable and interconnected societies. As technology continues to advance, the awareness of Geo-AI potential will only increase. The potential that artificial intelligence have when combining the capacity and the potential of GIS technologies is unbelievable. Being able to further analyse the spatial data and make better decisions, while not being limited by the opportunities that such analysis provide, leads to disruption. This combination further opens new opportunities for innovation.

## References


1. Arun S (2017) Digital twin—an emerging technology for geospatial world. Satpalda Geospatial Services. https://www.satpalda.com/blogs/digital-twin-an-emerging-technology-for-geospatial-world
2. Canzoneri M, de Luca A, Harttung J (2021) Digital twins: a general overview of the biopharma industry. Adv Biochem Eng Biotechnol 177:167–184
3. Garg PK (2019) Theory and principles of geoinformatics, Khanna Book Publishing. Co., Ltd., New Delhi
4. Garg and Arti (2025, in Press) An Introduction to LiDAR remote sensing. BSP Publication
5. Grieves M (2014) Digital twin: manufacturing excellence through virtual factory replication (Digital Twin White Paper-2014)
6. Excellence_through_Virtual_Factory_Replication
7. https://www.fortunebusinessinsights.com/digital-twin-market-106246
8. https://www.industryarc.com/Report/18531/geospatial-analytics-artificial-intelligence-market-research-report.html

9. Hu J, Lv ZL, Yuan D, He B, Chen W, Ye X, Li D, Yang G (2023), A spatiotemporal intelligent framework and experimental platform for urban digital twins. In: Virtual reality & intelligent hardware, vol 5, Issue 3, pp 213–231. https://doi.org/10.1016/j.vrih.2022.08.018
10. Kamel Boulos MN, Zhang P (2021) Digital twins: from personalised medicine to precision public health. J Pers Med 11:745. https://doi.org/10.3390/jpm11080745
11. Mai G, Xie Y, Jia X, Lao N, Rao J, Zhu Q, Liu Z, Chiang Y-Y, Jiao J (2025) Towards the next generation of geospatial artificial intelligence. Int J Appl Earth Observ Geoinform 136. https://doi.org/10.1016/j.jag.2025.104368
12. Mostak T (2021) The future of geospatial analytics is in digital twins. Forbes. https://www.forbes.com/sites/forbestechcouncil/2021/11/18/the-future-of-geospatial-analytics-is-in-digital-twins/?sh=168ffc7dd590.
13. Park J, Yang B (2020) GIS-enabled digital twin system for sustainable evaluation of carbon emissions: a case study of Jeonju City, South Korea. Sustainability 12:9186;https://doi.org/10.3390/su12219186
14. Stevens L (2019) Geospatial digital twins. https://medium.com/@linda_29745/geospatial-digital-twins-a8e2bc886fd4
15. Sharma L, Garg PK (eds) (2025) Digital twins and simulation technology concepts and applications, 1st edn. Chapman and Hall/CRC. https://doi.org/10.1201/9781003582489
16. Sharma L, Garg PK (eds) (2024) Deep learning in Internet of Things for next generation healthcare, 1st edn. Chapman and Hall/CRC. https://doi.org/10.1201/9781003451846
17. Sharma L, Garg PK (eds) (2023) Technological prospects and social applications of society 5.0, 1st edn. Chapman and Hall/CRC. https://doi.org/10.1201/9781003324720
18. Sharma L, Carpenter M (eds) (2022) Computer vision and Internet of Things: technologies and applications, 1st edn. Chapman and Hall/CRC. https://doi.org/10.1201/9781003244165
19. Sharma L, Garg PK (eds) (2021) Artificial intelligence: technologies, applications, and challenges, 1st edn. Chapman and Hall/CRC. https://doi.org/10.1201/9781003140351
20. Sharma L (ed) (2020) Towards smart world: homes to cities using Internet of Things, 1st ed. Chapman and Hall/CRC. https://doi.org/10.1201/9781003056751
21. Sharma L, Garg PK (eds) (2019) From visual surveillance to Internet of Things: technology and applications, 1st edn. Chapman and Hall/CRC. https://doi.org/10.1201/9780429297922
22. Sharma L (2019) Object detection with background subtraction. LAP LAMBERT Academic Publishing, SIA OmniScriptum Publishing, Latvia, European Union, February 2019. ISBN: 978-613-7-34386-9







# Chapter 5
# Digital Twin in Geospatial Science for Urban Planning and Development

**Shivam Kumar Gupta**

## 1 Introduction

Imagine a city where every street, building, utility, and tree or any object exists not only in the physical world but also within a real-time digital replica, an intelligent, smart, living digital model that grows, adapts, and evolves in lockstep with its physical counterpart. This vision represents the convergence of three powerful paradigms:

- The Digital City,
- The Smart City, and
- Responsive Urban Planning.

Together, they all form an ecosystem where technology, infrastructure, and society are interconnected to foster a more efficient, sustainable, safe, and citizen-centric urban future. At the core of this future city is geospatial science, acting as the spatial framework upon which all other technologies operate. It enables the precise mapping of city assets and dynamics, supporting real-time data integration, simulation, and analysis.

Suppose a city is considering building a new public park. The Digital Twin allows planners to simulate its impact on local air quality, social cohesion, traffic congestion, and property values before the first tree is planted. Similarly, adding a new metro line is no longer a matter of chance. The urban managers can predict commuter patterns, peak rush hours, energy use, and land value shifts, ensuring holistic and optimized outcomes and decongest the specific geographical area. Smart City functions as the nervous system of this urban ecosystem. Traffic signals are no longer static; they respond in real time to pedestrian and vehicle movement. Public transport lines adapt dynamically according to commuter demand and congestion conditions.

S. K. Gupta (✉)
Leica Geosystems Technologies, Singapore, Singapore
e-mail: shivamgupta@shivamgupta.net

 
L. Sharma et al. (eds.), *Digital Twin Technology in Geospatial Science: Trends, Applications and Challenges*, Data-Intensive Research,
https://doi.org/10.1007/978-3-032-20311-3_5

Intelligent waste management systems notify collection personnel when receptacles are full, minimizing superfluous trips and emissions. Energy consumption in buildings is autonomously regulated by AI-driven and trained algorithms that account for usage patterns, occupancy levels, and meteorological predictions. All operations are visible to administrators and available to citizens via digital interfaces, enhancing governance transparency and participation.

Urban planning in a smart city is no longer limited to fixed blueprints and inflexible five-year strategies. Planners utilize dashboards to illustrate economic activity, social behavior, people's safety, and environmental changes. Should a technology park in the city draw a substantial workforce, the system predicts high residential demand in the vicinity, notifies housing developers, and initiates enhancements in water and transportation infrastructure. Policymakers engage with residents via immersive digital platforms that enable virtual exploration and voting on proposed zoning modifications or infrastructure initiatives. Residents transform into co-creators of their surroundings rather than only passive recipients.

This integration of the digital and physical realms allows cities to progress seamlessly. Urban expansion evolves to be modular and efficient, directed by simulations and feedback mechanisms. New districts are established in response to real-time demographic changes and climate forecasts. Established neighborhoods are enhanced with intelligent and smart technologies and sustainable infrastructure, augmenting and creating quality of life and resilience. Urban greenery is meticulously tuned using models that assess the influence of tree canopies on air quality, noise levels, and temperature. The strength of this city resides in its inclusivity and sustainability. Emergency services execute simulated drills utilizing the city's Digital Twin to improve evacuation strategies, resource allocation, and reduce response time. In the event of a genuine crisis, emergency teams have real-time updates and operate effectively, minimizing casualties and damage.

The city prioritizes its residents above all else. Services are tailored, adaptive, and comprehensive. Citizens obtain prompt information regarding services, can report difficulties immediately via smart applications, and are engaged participants in driving governance. Senior citizens encounter sidewalks intended for their mobility; youngsters attend schools strategically located according to demographic projections; firms have legislative assistance connected with digital advancements. Technology serves as a facilitator of well-being, rather than only enhancing efficiency.

This chapter talks about how Digital Twins are becoming more important in urban management and how they are becoming more common in geospatial science. We will show you how this convergence of technologies is making cities more resilient, responsive, and focused on the needs of their citizens through real-world examples, use cases, and analysis. The chapter aims to provide a theoretical framework and a practical guide for urban planners, technologists, and policymakers to utilize Digital Twins in sustainable urban development.

## 2 The Cognitive Layer: Digital Twin, AI, and Urban Intelligence

In the swiftly urbanizing world, cities confront escalating challenges, including infrastructure pressure, environmental degradation, mobility issues, and increasing socio-economic inequality. Urban planners face escalating pressure not only to adapt to these changes but also to foresee and manage them proactively. This increasing complexity necessitates a fundamental change in the conceptualization, planning, and management of cities. The combination of Digital Twin technology with Geospatial Science is leading this revolution, providing a dynamic, data-driven methodology for urban planning and development.

A Digital Twin is a real-time, digital representation of physical systems, surroundings, or infrastructure. When combined with geographical data—sourced from satellites, sensors, aerial images, and GIS platforms—it establishes a robust simulation environment in which cities may be visualized, analyzed, and optimized unprecedentedly. This integration enables urban planners to model hypothetical situations, assess infrastructure integrity, predict urban expansion, manage emergencies, and optimize resource utilization, all within a geographically precise virtual setting.

Geospatial Digital Twins are not simply digital maps or 3D models; they are sophisticated ecosystems that embody the physical, functional, and behavioral characteristics of urban systems. They integrate real-time data, historical patterns, citizen behavior, and environmental variables to facilitate both rapid and strategic decision-making. Digital Twins are transforming the future of urban design through intelligent energy networks, autonomous transportation systems, sustainable infrastructure, and accessible public areas.

## 3 Conceptual Foundation of a Digital Twin

A Digital Twin is a virtual replica of a physical object, system, or environment that mirrors its real-world counterpart in real time. It combines data from sensors, IoT devices, historical records, and simulations to create a dynamic model that updates and evolves as the actual physical asset changes. In simple terms, it's like having a live, intelligent "mirror" of something physical, such as a building, a transportation network, or even an entire city, that you can observe, analyze, test, and improve without interfering with the real-world version.

Key characteristics of a Digital Twin include [2].

- Real-Time Data Integration: It receives continuous input from physical sensors and systems.
- Simulation Capability: It allows testing of different scenarios (e.g., infrastructure stress during a flood).
- Predictive Analytics: It uses AI and machine learning to forecast behaviors and outcomes.

**Fig. 1** The ecosystem of a Digital Twin [3]. *Source* Hexagon report

- Visualization Tools: Often includes 3D models, dashboards, software & solutions, and immersive environments.

A Digital Twin becomes even more powerful when tied to geospatial data, mapping the digital model onto the real-world geography. This enables urban planners to simulate traffic, analyze land use, monitor environmental changes, or plan new developments with spatial accuracy and context.

The ecosystem of a Digital Twin is presented in Fig. 1.

## 4 Components of a Digital Twin

A typical Digital Twin consists of three entities as shown in Fig. 2.

**Fig. 2** Components of a digital twin

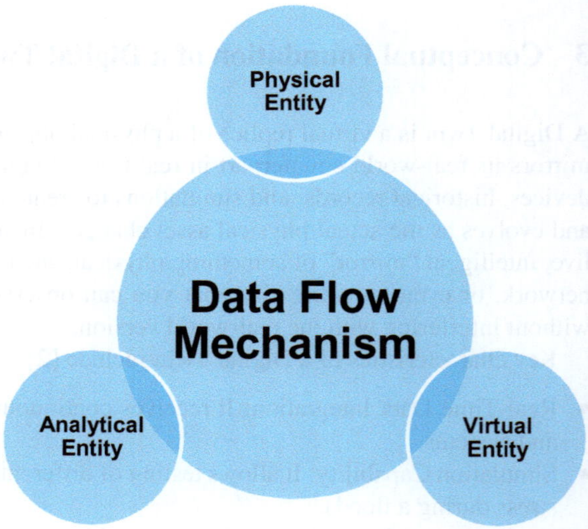

- Physical Entity: The real-world system being replicated.
- Virtual Entity: The digital counterpart that simulates the behavior of the physical entity.
- Analytical Entity: It enables proactive planning, sustainable development, and efficient resource use by turning data into foresight and relevant information.

Data Flow Mechanism: This is an invisible glue that keeps the different entities together. This is a Bi-directional data exchange for real-time synchronization among all three entities.

## 4.1 Categorization of Digital Twins: A Multi-Dimensional Perspective

Digital Twin technology is no longer confined to a single domain or form. With increasing integration into urban environments, it becomes essential to understand Digital Twins not just in terms of components or assets, but through a broader lens that includes their scope and scale, functional maturity, and technological depth. The stakeholders align the deployment of Digital Twin technology with the desired outcomes in urban planning, infrastructure optimization, and citizen services based on the categorization mentioned below.

## 4.2 Categorization by Scope and Scale

Digital Twins vary significantly in their scale from micro-level assets to entire ecosystems, as given in Table 1. Categorizing them by scope allows city planners and technologists to modularize implementation and integrate systems incrementally [2].

## 4.3 Categorization by Functional Capabilities

Digital Twins also evolve based on what they are designed to do (Refer to Table 2). This functional classification emphasizes the progression from simple data representation to predictive and autonomous systems [4].

**Table 1** Digital twins as per their scale

| Type | Description | Example |
|---|---|---|
| Micro-level twins | These include Component and Asset Twins that model small-scale elements like machinery, lighting poles, HVAC systems, or buildings. They serve as the foundational blocks for larger twins and are ideal for condition monitoring, maintenance, and early anomaly detection | A digital twin of a traffic signal junction in Los Angeles continuously assesses performance to reduce idle time and carbon emissions |
| Meso-level twins | Meso-level or System Twins represent interconnected networks such as transportation corridors, utility grids, or neighborhoods. These help in planning coordinated responses, optimizing performance across connected assets, and managing urban subsystems cohesively | Amsterdam's mobility Digital Twin integrates bike lanes, public transit, and pedestrian pathways to manage flow and enhance safety |
| Macro-level twins | These are city-scale or regional twins encompassing an entire urban ecosystem. They include infrastructure, social dynamics, environment, and governance layers, allowing long-term scenario planning, policy modeling, and climate impact analysis | "Virtual Singapore" exemplifies a macro-level twin where multiple urban functions and agencies work within a shared 3D digital environment |

**Table 2** Digital twins based on their design

| Type | Description | Use case |
|---|---|---|
| Descriptive twins | These offer real-time or near-real-time visualization of a physical system. They serve as digital mirrors but don't conduct analysis. Useful for monitoring and diagnostics | Monitoring energy consumption in a smart building dashboard |
| Diagnostic twins | These analyze data to understand why a situation has occurred. By incorporating historical and contextual data, diagnostic twins assist in root cause analysis | A utility network twin identifying reasons behind repeated water leakages in a certain district |
| Predictive twins | These go a step further by using simulation, machine learning, or forecasting algorithms to anticipate what could happen. They are essential for proactive urban management | Predicting flood zones in real time using rainfall and drainage data in Mumbai |
| Prescriptive twins | These recommend or automate decisions based on predictive insights. They enable what-if analysis and policy testing, supporting dynamic urban governance | A transportation twin adjusting traffic light sequences automatically based on congestion trends |
| Cognitive or autonomous twins | The most advanced category, these use AI to self-learn and adapt without human intervention. They are future-facing models aimed at autonomous city systems | A digital twin of an autonomous port that schedules ship arrivals, resource allocation, and maintenance based on AI inputs |

### 4.4 Categorization by Technology Integration

Digital Twins can also be differentiated based on the maturity and complexity of technologies they integrate, as given in Table 3. This classification is crucial for IT architects and urban tech consultants.

The landscape of Digital Twins is diverse and rapidly evolving. By categorizing these systems through the lenses of scope, functionality, capability, volume, and technological integration, urban stakeholders can adopt a layered, phased, and strategic approach to implementation. Whether monitoring a single asset or orchestrating the operations of a megacity, the appropriate Digital Twin category determines the value and impact it delivers. Therefore, it is very important to understand the requirements and deliverables users are looking at and focusing on which applications. As urban environments face increasing pressure from climate change, population growth, and infrastructure demand, these categorizations will provide scalable, smart, and sustainable solutions tailored to each city's unique context, culture, and community.

## 5 Geospatial Science and Digital Twins

The fusion of Geospatial Science with Digital Twin technology is reshaping the way modern cities are designed, managed, and transformed. As urban environments become more complex, the need for tools that combine spatial awareness with real-time intelligence becomes critical. Geospatial science, with its ability to capture,

**Table 3** Digital Twins based on the maturity and complexity of technologies

| Type | Description | Example |
|---|---|---|
| Basic twins | Use static 3D or GIS data with minimal sensor input. Often used in early-stage planning or legacy infrastructure mapping | A basic city zoning map with overlays of demographic data for land-use planning |
| Connected twins | These integrate real-time data from IoT devices, weather stations, or mobile sensors. They allow for dynamic updates and better decision support | Smart water grids in Seoul that adjust pressure and flow based on real-time consumption patterns |
| Smart twins | Combine AI/ML algorithms for anomaly detection, forecasting, and optimization. These twins drive actionable intelligence across city operations | A smart energy twin that learns from consumption patterns and auto-regulates district heating in winter |
| Intelligent twins | These merge edge computing, cloud processing, blockchain (for secure transactions), and immersive tech (AR/VR) to create decentralized, intelligent ecosystems | A twin enabling virtual town halls using AR overlays of city development proposals for community input in real time |

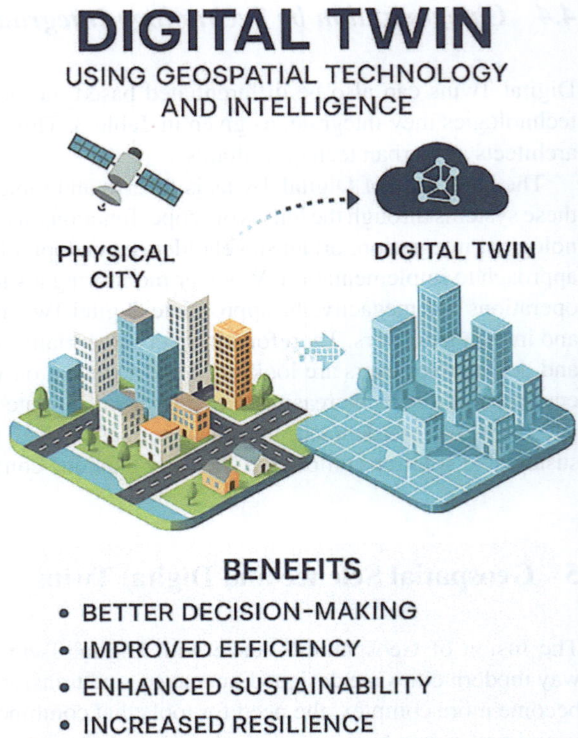


**Fig. 3** Geospatial technology and geospatial intelligence

analyze, and visualize data linked to specific locations, serves as the spatial backbone of the Digital Twin, as shown in Fig. 3. When integrated seamlessly, these technologies empower planners and policymakers to create responsive, efficient, resilient urban systems and greater sustainability.

- **Real-Time Urban Monitoring and Spatial Awareness**: One of the major benefits of integrating geospatial data with Digital Twins is the ability to monitor urban areas in real time. Inputs from satellites, satellite imagery, aerial surveys, drones, and ground-based sensors for ground truthing are used to inform digital models, enabling town planners and city managers to track conditions such as traffic congestion, peak and non-peak zones, air pollution, noise levels, or construction progress and take data-driven, informed decisions.

  *For example, in Amsterdam, geospatial-enabled twins track cycling and pedestrian movement patterns throughout the day, helping city officials identify congested zones and plan safe, efficient mobility routes. This real-time awareness transforms passive infrastructure into active, responsive systems used for decongestion and optimization of routes.*

- **Simulation and Predictive Analysis for Urban Risk**: Geospatial science plays an important role in enhancing the predictive power of Digital Twins by adding location-based data into simulations. Urban planners can predict the effects of flood, heatwave, or earthquake-prone zones on neighborhoods by integrating spatial layers, including terrain, land cover, infrastructure, and population density. These models facilitate the development of more resilient emergency response techniques, procedures, and informed long-term climate adaptation planning.

  *Another example is consider Tokyo, where Digital Twins replicate seismic activity with geographical terrain and fault line data, thereby facilitating the city's development of more efficient evacuation procedures and the fortification of essential infrastructure in vulnerable regions.*
- **Land Use Optimization and Smart Zoning**: Digital Twins integrated with geospatial data will help in transforming land use planning by offering a multi-dimensional perspective of urban environments. The integration of land registration data, property boundaries, zoning regulations, and topographical models facilitates more intelligent land resource allocation. Authorities can evaluate project viability, recognize regulatory limitations, reduce downtime, and anticipate future developmental effects.

  *In Seoul, geospatial digital models assess sunlight access, wind flow, and population distribution before the approval of high-rise building constructions, ensuring that environmental and social factors are well considered in planning choices.*
- **Revolutionizing Transportation Intelligence**: The combination of Geospatial Science, Geo-Intelligence and Digital Twin Technology is transforming the surface transport and railway industry. These tools enhance current performance and facilitating the scope for future development of Smart & Safe Railways, encompassing strategic planning, real-time operations, and sustainability. Governments, transportation authorities, and commercial operators are investing in this synergy to establish a robust, data-driven, and safe transportation infrastructure.

  *The UK's HS2 incorporates 3D geospatial data with real-time integration of twin updates to track construction progress and environmental impact; Deutsche Bahn merges IoT-driven twins with GIS to monitor assets, predict asset failures, and enhance punctuality; and Indian Railways employs drone imagery, soil data, and virtual modeling to effectively modernize rural routes, based on passenger density and seasonality's.*
- **Sustainability and Environmental Stewardship**: As urban areas strive to achieve climate objectives, Digital Twins integrated with geospatial intelligence provide precise actions for environmental sustainability. Spatial data is essential for environmental modeling, whether for improving green space distribution, assessing urban heat islands, or regulating stormwater flow.

  *Vienna utilize GIS technology for its urban twin to analyze energy consumption trends, green coverage, and emissions data, facilitating the strategic installation of rooftop solar panels and tree planting projects, help in advancing its climate neutrality objectives.*

- **Citizen-Centric Spatial Engagement**: Digital Twins, when coupled with geospatial systems, augment public involvement. Citizens can engage with 3D maps of their areas, examine proposed infrastructure modifications, and comprehend how regulations may affect their daily lives. This interactive method enhances transparency and promotes cooperative urban development.

  *Virtual Singapore, a premier program, enables both government agencies and residents to engage with a meticulously detailed 3D geospatial model of the city, supporting planning debates, submission of comments, and visualization of policy consequences—a Reality check.*

# 6 Technologies Enabling Digital Twins

The effectiveness of a Digital Twin relies on a sophisticated ecosystem of technologies that work hand in hand to mirror the physical world in real-time. Right from the collection of high-fidelity data to intelligent analysis and immersive visualization, a range of cutting-edge technologies is imperative for the creation, operation, and evolution of Digital Twins [5]. It is the fusion of different sensors, software's, geospatial platforms, IoT devices, cloud computing, artificial intelligence (AI), and advanced simulation tools. Together, they empower not only the replication of physical assets but also the predictive, prescriptive, and interactive capabilities that make Digital Twins transformative for urban planning and development, as presented in Table 4.

## *6.1 Geographic Information Systems (GIS)*

GIS are great for collecting, storing, analyzing, and showing location, and geographic data. When GIS is integrated with Digital Twin technology, it makes it much easier to accurately, efficiently, and intelligently simulate real-world settings due to its layered structure. In a Digital Twin environment, GIS serves as the base layer, providing actual and virtual objects that are being mimicked in a geospatial context. For applications like disaster management, traffic flow optimization, and sustainable land use, the synergy between GIS and Digital Twins is very important. It enables cities to plan better and respond quickly to changes.

## *6.2 Remote Sensing*

Remote sensing, via satellites, aerial images, high-resolution satellite imagery, digital images, and unmanned aerial vehicles (UAVs), is essential for Digital Twins because

**Table 4** Technologies employed for digital twins and their principal functions

| Technology | Primary function | Use cases |
|---|---|---|
| IoT | Real-time data collection | Monitoring bridge stress in smart infrastructure |
| GIS and geospatial | Spatial context and mapping | Land use planning in urban development |
| BIM | Detailed infrastructure modeling | Optimizing hospital layout and energy usage |
| AI & ML | Prediction and automation | Forecasting traffic patterns |
| Cloud and edge computing | Storage, computation, and real-time processing | Processing air quality data at the stress level |
| 3D/AR/VR visualization | Interactive interfaces for users | Immersive planning of new public parks |
| Big data analytics | Insight generation from diverse datasets | Energy consumption trends for city districts |
| Cyber-physical systems | Feedback loops between physical and digital systems | Smart streetlights adjusting to foot traffic |
| APIs & middleware | System integration and interoperability | Combining transit and weather data in a city twin |
| Blockchain | Data integrity and secure collaboration | Verified asset maintenance logs |

it provides up-to-date, consistent, and high-resolution geographic data. This information makes it possible to keep an eye on how land is being used, the state of the environment, and infrastructure improvements in both urban and rural areas.

In the Digital Twin ecosystem, remote sensing is a powerful "eye in the sky" that ensures the digital model remains up to date with what is happening in the real world. Real-time or periodic updates to remote sensing improve the Digital Twin's temporal accuracy and efficiency, which makes simulations and treatments more effective. It is a basic technology that makes Digital Twins work and grow in cities. Applications include.

- Detecting urban sprawl and illegal encroachments,
- Monitoring construction progress and regulatory compliance,
- Mapping green cover changes and the heat island effect, and
- Supporting flood modeling and disaster response planning.

## *6.3 Internet of Things (IoT)*

The various IoT devices monitor the data collected from traffic, pollution, energy use, and the health of buildings and critical structures in real time. These sensors send data to the digital twin all the time 24 × 7, so that it may be updated in real time. The IoT is the central nervous system of a Digital Twin, supplying it with the real-time, high-frequency, and multi-modal data required to maintain a living, breathing representation of the physical urban environment.

1. Real-Time Data Collection and Synchronization: IoT devices, such as sensors, actuators, smart meters, and embedded controllers, continuously capture data from:

   - Traffic flow and congestion levels,
   - Air and noise pollution,
   - Water and energy consumption,
   - Waste management systems, and
   - Public transportation and pedestrian movement.

This data is streamed into the Digital Twin platform via cloud gateways or edge devices, enabling the synchronization of the virtual model with physical reality.

2. Enabling Real-Time Simulations and Interventions: With live data from IoT devices, Digital Twins can simulate and evaluate scenarios such as

   - Optimizing traffic signal timing based on congestion data,
   - Predicting and pre-emptively mitigating energy overloads,
   - Adjusting HVAC systems in smart buildings based on occupancy, and
   - Automatically rerouting buses in response to road blockages.

IoT thus transitions the Digital Twin from a diagnostic tool to a proactive decision engine.

3. Infrastructure and Asset Monitoring: Smart infrastructure is embedded with sensors that detect:

   - Structural stress and vibration in bridges and buildings,
   - Leakages in water or gas pipelines,
   - Load levels in electric grids, and
   - Elevator performance and failure diagnostics in high-rises.

This capability enables predictive maintenance, reducing costs and preventing disasters. The Digital Twin constantly updates itself to compare the current health of urban assets with the previous version.

4. Urban Governance and Citizen Engagement: Downstream IoT platforms, like smart streetlights, parking sensors, and waste bins, interact directly with the citizens and help in their daily lives. These systems improve service delivery, safety, smart living, and feed the Digital Twin with citizen-driven data, allowing:

   - Participatory planning,
   - Demand-responsive public transport, and
   - Real-time feedback loops between residents and authorities.

This democratization of data through IoT fosters inclusive urban governance.

## 6.4 Building Information Modeling (BIM)

The growth driver of Digital Twin technology in urban planning has been profoundly influenced by progress in 3D city modeling, fueled by point clouds, Building Information Modeling (BIM), and semantic data models such as CityGML. These technologies collectively furnish the geometric, semantic, and functional framework essential for creating high-fidelity, responsive digital copies of actual urban environments.

3D city modeling comprises of generating detailed, spatially accurate depiction of urban structures and terrains. These models represent visualization and analyzation by integrating geometric accuracy, structural hierarchies, and contextual interactions among urban features. Within Digital Twin architecture, 3D models are agile and dynamic; they are interconnected with the help of real-time data streams from IoT sensors, environmental systems, and infrastructure networks. This combination enables planners to model urban processes, examine spatial patterns, and facilitate decision-making with significant realism and contextual relevance.

BIM is a technique of representing physical and functional characteristics of any physical object, here we take example of buildings. It emphasizes digital processes for the design, construction, and management of building assets utilizing comprehensive information models. Every BIM model has not just 3D geometry but also metadata regarding materials, lifecycle phases, energy use, and maintenance records. The integration of BIM into a city-scale Digital Twin facilitates micro-level examination of individual structures within the broader urban environment. This facilitates extensive simulations, including the influence of a building's energy performance on district-level power consumption and the effects of refurbishment on pedestrian or traffic patterns. A city planning authority may utilize BIM-integrated Digital Twins to model the effects of proposed high-rise constructions on high wind-prone corridors, daylight exposure, and emergency access routes. These elements are essential for resilient and sustainable urban development.

Moreover, CityGML (City Geography Markup Language) is an open, standardized data model and XML-based format intended for the storage and exchange of three-dimensional city models. It transcends basic geometry by incorporating semantic information, including building types, functions, construction materials, and legal borders, which are crucial for substantive analysis and interoperability. In a Digital Twin environment, CityGML facilitates the organized depiction of urban components at several levels of detail (LoDs). This semantic richness enables diverse systems, like GIS platforms, BIM tools, and simulation engines, to communicate efficiently, allowing consistent knowledge and analysis across various planning and operational domains. CityGML is particularly advantageous for city-wide Digital Twins necessitating multi-scale representation, encompassing individual buildings, entire neighborhoods, and infrastructure systems.

The integration of 3D city modeling, BIM, and CityGML within Digital Twins enables cities to transition from reactive planning to predictive, data-informed governance. Planners can model prospective growth scenarios, evaluate environmental and social repercussions, and examine infrastructure resilience under stress events such as flooding, heatwaves, or traffic surges. Furthermore, these technologies facilitate collaborative engagement among stakeholders—such as architects, engineers, policymakers, and citizens, helping them in the planning process via interactive visualizations and simulation tools.

## 6.5 Cloud Computing and Big Data Analytics

Cloud computing and big data analytics serve as essential facilitators for the scalability and intelligence of Digital Twin systems. Utilizing cloud platforms, Digital Twins can function in real-time across geographically dispersed systems, providing uninterrupted access to data and computing resources for planners, engineers, and decision-makers.

Cloud computing and big data analytics collectively constitute the digital foundation that converts static urban models into intelligent, adaptive, and scalable Digital Twin ecosystems. These insights enable Digital Twins to model prospective scenarios, enhance performance, and facilitate predictive urban planning. By monitoring mobility data via cloud-based systems, local officials can dynamically modify traffic signals, anticipate congestion hotspots, or redirect public transportation to enhance flow and reduce emissions.

**Cloud Computing and Big Data Analytics**

Cloud computing offers the adaptable, on-demand infrastructure necessary to host and handle the extensive data produced by sensors, IoT devices, GIS platforms, and simulation models in urban settings. As per the requirement and specifications, developers can use on-premises, hybrid, or fully cloud deployment methodology. In addition, big data analytics examine vast and varied datasets, including traffic patterns, energy use, environmental factors, and social behaviors, to reveal concealed patterns, trends, and anomalies.

## 6.6 Artificial Intelligence and Machine Learning

Artificial Intelligence (AI) and Machine Learning (ML) significantly enhance the predictive and adaptive functionalities of Digital Twins. Artificial Intelligence (AI) focuses on emulating human cognitive functions by machines, whereas Machine Learning (ML) is a subset of AI that enables systems to discern patterns in data and enhance decision-making autonomously over time, without explicit programming. In the realm of Digital Twins, these technologies facilitate the ongoing analysis and interpretation of extensive, real-time datasets generated by urban sensors, infrastructure networks, and citizen platforms.

AI-driven Digital Twins can independently model complex urban phenomena, such as traffic congestion, energy utilization, or flood hazards, and propose effective remedies and modification required for future planning. Machine learning models, utilizing historical and real-time data, improve forecast and prediction accuracy, identify abnormalities, and automate responses. In the context of the smart city, machine learning algorithms integrated within a Digital Twin might predict surges in power demand and initiate energy conservation measures or redirect traffic because of unforeseen disturbances. This synergy converts Digital Twins from static replicas into intelligent, self-learning systems that facilitate proactive urban planning and robust city management.

Digital Twins, enhanced by geospatial science, data intelligence, and AI, offer a comprehensive, systems-oriented perspective of the urban landscape. They enable cities to diminish their carbon footprint via enhanced design, optimized operations, and adaptive planning. As urban difficulties intensify, these digital ecosystems will be essential instruments in attaining global climate objectives and fostering a planet where urban expansion and sustainability coexist harmoniously.

## 7 Case Studies

### *7.1 Singapore: Virtual Singapore [7]*

Virtual Singapore represents a cutting-edge Digital Twin initiative, providing a three-dimensional city model for urban planning, modification, environmental simulations, and public involvement. The platform consolidates real-time data from multiple sectors and facilitates inter-agency collaboration.

### *7.2 Helsinki, Finland [8]*

Helsinki has established a comprehensive Digital Twin for the analysis and monitoring of carbon emissions, traffic optimization, and citizen engagement. The initiative promotes sustainable urban development objectives.

### *7.3 Shanghai, China [9]*

Shanghai's Digital Twin initiative facilitates real-time surveillance and monitoring of metropolitan infrastructure, critical structures, transportation networks, and environmental parameters. It demonstrates how high-density urban areas might utilize digital twins for effective governance.

### 7.4 India: Amaravati Smart City [10]

Despite being in preliminary phase, Amaravati's Digital Twin efforts aim to integrate GIS, IoT, and BIM for urban design, infrastructure oversight, and citizen services.

## 8 Geospatial Technology for Sustainability

### 8.1 Cities and Climate in the Digital Era

With the rapid acceleration of global urbanization, cities are becoming as the focal points of economic activity and environmental strain. Urban regions contribute to more than 70% of worldwide $CO_2$ emissions, primarily attributable to traffic, construction, and energy consumption.

*"by 2050, the world's population is set to increase to 9.9 billion" Source: UN*

In response, urban planners are progressively adopting Digital Twin technologies combined with geospatial science to create more intelligent, sustainable, and efficient urban spaces. Digital Twins provide an effective framework for modeling and regulating the urban ecosystem, thereby facilitating sustainability and climate objectives.

### 8.2 Visualizing Environmental Impact in Real Time

Digital Twins facilitate urban decision-makers in visualizing and evaluating the environmental consequences of infrastructure projects prior to their construction by generating a real-time, data-driven representation of a city. These models replicate variables including air quality, traffic emissions, heat island phenomena, and energy consumption patterns. In Melbourne, Australia, urban planners used a Digital Twin to assess how tree canopy and reflective building materials may decrease metropolitan heat by as much as 4 °C, thereby reducing energy usage and enhancing resident comfort.

### 8.3 Optimizing Transportation for Lower Emissions

Transportation systems significantly contribute to urban carbon footprints. Digital Twins combined with geospatial traffic and mobility data help cities to model and enhance their transportation networks. A mobility-centric Digital Twin in Rotterdam, Netherlands, perpetually simulates the flows of bicycles, pedestrians, and vehicles. This enables the city to modify traffic signals instantaneously, alleviate congestion, and promote more sustainable commute alternatives. Likewise, public transportation can be optimized according to anticipated demand, thereby decreasing fuel consumption and enhancing service efficiency.

### 8.4 *Energy-Efficient Urban Design*

The construction and operation of structures substantially contribute to greenhouse gas emissions. Digital Twins, in conjunction with BIM and GIS, enable planners to model the energy efficiency of buildings, neighborhoods, or entire districts. Vienna has created an Urban Energy Twin to simulate the energy consumption and heating needs of its buildings. This methodology directs retrofitting initiatives and facilitates the incorporation of renewable energy sources, assisting Vienna in achieving its carbon reduction objectives.

### 8.5 *Green Infrastructure and Nature-Based Solutions*

Geospatially enabled Digital Twins can model the effects of nature-based initiatives, including green roofs, urban trees, and stormwater management systems. These green infrastructures augment urban biodiversity while also mitigating carbon emissions through pollution absorption and temperature reduction. For example, Singapore's Virtual Singapore initiative encompasses simulations for tree planting and building shading to diminish energy use in residential areas while enhancing the urban microclimate.

### 8.6 *Waste and Water Management for Sustainability*

Effective management of waste and water systems is a crucial factor in diminishing urban carbon footprints. By integrating sensor data, previous trends, and spatial analytics into a Digital Twin. The urban areas can monitor water leaks, identify inefficiencies in sewage systems, and optimize garbage collection routes. This conserves resources and reduces the emissions linked to these systems. Barcelona employs intelligent waste collection and routing to reduce fuel consumption by municipal trucks, thereby immediately decreasing emissions.

## *8.7 Predictive Capabilities for Climate Resilience*

Digital Twins serve as a preventive function by simulating climate-related hazards, including flooding, droughts, and energy shortages. These prognostic insights inform infrastructure adaptation, zoning decisions, and emergency response methods. New York City's FloodNet amalgamates flood sensors with Digital Twin models to deliver early warnings and longitudinal planning data, facilitating more intelligent stormwater management and resilience solutions in at-risk communities.

## *8.8 Urban Planning Applications*

1. Land Use and Zone Management: Digital Twins enable city/town planners to visualize existing land use patterns and model zoning modifications. They can also conduct an impact evaluation before implementation, therefore improving decision-making.
2. Infrastructure Management and Monitoring: Utilities, including water supply, energy, and waste management, can be monitored and optimized using Digital Twins. The timely maintenance schedules, load management, and failure predictions attain enhanced precision.
3. Safe Transportation & Mobility: The Digital Twins simulate traffic patterns, public transit systems, and pedestrian behavior. They provide the simulation of policy modifications, such congestion pricing and the implementation of bike lanes, while evaluating their real-time effects.
4. Environmental Monitoring: Digital Twins provide an assessment of air and water quality, urban heat islands, heat maps, and vegetation coverage. They advocate for the development of sustainable urban settings via data-informed policies.
5. Urban Expansion and Scenario Planning: Scenario analysis utilizing Digital Twins enables cities to model growth trajectories, climatic effects, and socio-economic transformations. This facilitates long-term strategic planning.
6. Disaster Management: The Digital Twins are essential in emergency preparedness and catastrophe response. They provide the simulation of flood hazards, fire propagation, and evacuation techniques, thereby enhancing preparedness and resilience.

# 9 Future Prospects and Research Directions

## *9.1 Integration with AR/VR and the Metaverse*

The integration of Digital Twin technologies with Augmented Reality (AR), Virtual Reality (VR), and the expanding Metaverse signifies a significant transformation in the design, experience, and management of urban areas. Augmented reality superimposes real-time data onto the physical environment, enabling planners and engineers to examine subterranean utilities, traffic patterns, or infrastructure issues on-site through mobile devices or smart glasses [15–18]. Conversely, VR generates comprehensive simulations enabling stakeholders to examine and engage with a complete virtual depiction of a city, whether for public discussions, design validation, or emergency response training. The Metaverse enhances this by facilitating a continuous, collaborative digital environment where urban planners, residents, and enterprises can jointly build and model urban development in real-time.

As Digital Twins advance, their incorporation with AR/VR and Metaverse platforms will transform citizen participation, participatory planning, and infrastructure lifecycle management. A digital reproduction of a city square in the Metaverse might facilitate virtual town hall meetings, allowing inhabitants to tour 3D environments, offer input, and propose alterations. Research is increasingly centered on facilitating real-time synchronization between physical assets and their immersive counterparts, enhancing spatial precision, interface fidelity, and data security. Digital Twins are expected to function as fundamental components of the Metaverse, enabling predictive modeling, intelligent automation, and democratized urban innovation. This convergence offers significant potential for developing transparent, inclusive, and adaptable cities that are more responsive to human and environmental requirements.

## *9.2 Citizen Engagement Platforms*

Citizen engagement platforms are becoming essential elements of urban Digital Twin ecosystems. It helps in facilitating direct communication between city planners and the citizens. These systems provide bidirectional communication, allowing users to receive urban information while also providing data, opinions, and ideas that impact planning decisions. When combined with Digital Twins, engagement systems enable citizens to observe proposed infrastructure projects, evaluate environmental implications, and engage in scenario simulations using user-friendly 3D models or interactive dashboards. A community redevelopment plan can be represented through a digital twin interface, allowing citizens to vote on design possibilities or propose improvements based on their experiences.

The future of citizen involvement will focus on fostering more inclusive, real-time, and immersive interactions. Advancements in mobile technology, augmented reality, and natural language processing are anticipated to transform platforms into intelligent dialogue systems that read feedback, customize urban services, and democratize decision-making. Research is progressively investigating how AI can assess public mood, prioritize concerns, and propose adaptive urban strategies. There is an increasing emphasis on creating gamified participation tools that promote broader engagement from many communities, especially youth and underrepresented populations. As Digital Twins evolve, their incorporation with citizen engagement platforms will transition city planning from a hierarchical paradigm to a collaborative, co-creative approach, allowing cities to become more adaptive, inclusive, and sustainable amid swift urban and digital transformation.

## 9.3 Autonomous Urban Systems

Autonomous urban systems signify the forthcoming advancement in smart city evolution, incorporating self-regulating infrastructures like traffic management, garbage collection, public transportation, and energy distribution. When integrated with Digital Twin technology, these systems acquire a dynamic, data-driven layer that facilitates continuous monitoring, modeling, and optimization of operations autonomously. A Digital Twin serves as a real-time control interface for autonomous operations, acquiring data from IoT sensors integrated into roads, buildings, or vehicles, and employing AI algorithms to make decisions, such as rerouting traffic in response to congestion or modulating energy flow based on demand variations.

The integration of Digital Twins with autonomous systems has significant potential for improving urban efficiency, resilience, and sustainability. Future research focuses on developing entirely integrated ecosystems in which autonomous subsystems, such as drone-based delivery, driverless public transit, autonomous vehicles, or AI-managed utilities, connect effortlessly over shared Digital Twin frameworks. There is increasing interest in creating self-learning models that evolve over time in response to environmental changes and citizen behavior. Moreover, ethical aspects including responsibility, privacy, and algorithmic transparency are becoming crucial study domains to guarantee responsible implementation. In the future, Digital Twins will be essential in managing autonomous urban environments, enhancing their adaptability, self-correction, and responsiveness to real-time citizen requirements.

## 9.4 Cross-Domain Integration

Cross-domain integration refers to the seamless collaboration of multiple urban systems, such as transportation, energy, water, environment, irrigation, and public safety, within a cohesive framework [13]. This integration, powered by Digital Twin technology, provides a holistic and integrated perspective of urban ecosystems, enabling decision-makers to comprehend the cascading impacts of alterations in one area on others. For instance, modifying traffic patterns according to real-time pollution metrics or enhancing building energy efficiency in accordance with meteorological conditions and grid capacity. A Digital Twin functions as a central nervous system that consolidates data across several domains and departments and facilitates coordinated simulations, predictive analytics, and comprehensive optimizations.

The future of cross-domain integration depends on the creation of interoperable digital twin systems that comply with common data standards, APIs, and ontologies. Recent research concentrates on developing federated digital twins. The modular models that represent distinct domains capable of secure and intelligent interaction and data exchange. Progress in edge computing, 5G connectivity, and semantic web technologies are anticipated to enhance real-time data sharing among systems. Moreover, machine learning is being used to uncover latent relationships and propose cross-sector actions. This comprehensive strategy will be crucial for addressing intricate urban issues such as climate resilience, catastrophe response, and transitions to a circular economy. Through the dismantling of silos and the promotion of collaboration, Digital Twin-enabled cross-domain integration will be essential in creating smarter, more adaptive, more systemically efficient future cities.

## 9.5 *Academic Research and Innovation*

Academic research and innovation are essential catalysts for the advancement of Digital Twin technology, particularly in urban planning and geospatial science [14]. Universities and research institutes are leading the investigation of innovative approaches to improve the precision, scalability, and efficacy of Digital Twins and their new applications in different industries. This entails the enhancement of algorithms for real-time data assimilation, the integration of diverse data sources, and the advancement of predictive analytics driven by artificial intelligence and machine learning. Research is advancing into hybrid models that integrate physics-based simulations with data-driven methodologies, resulting in more resilient and adaptable digital representations of intricate urban systems.

Future interdisciplinary collaboration across civil engineering, computer science, urban studies, environmental science, and social sciences is essential for realizing the complete potential of Digital Twins. The recent research initiatives emphasize the ethical use of AI, the safeguarding of privacy in data sharing, and the adoption of citizen-centric design principles that promote inclusivity and openness. Moreover, innovation centers and academic consortiums are utilizing Digital Twin testbeds and living labs to assess technologies in actual urban settings, hence expediting knowledge transfer to municipalities and the business sector [19–22]. As Digital Twins increasingly integrate into urban governance and policy-making, ongoing academic research will be essential to tackle challenges such as system interoperability, data security, and socio-technical implications, ensuring that the development of Digital Twins corresponds with sustainable and equitable urban futures.

## 10 Conclusion

The combination of digital twins with geospatial science provides a game-changing method for city planners and town developers. Planners, legislators, and stakeholders can now experience a improved level of visualization, simulation, and optimization of complicated city systems never possible before. Thanks to Digital Twins, which generate dynamic, real-time digital copies of actual urban settings. Thanks to this confluence, better data-driven decisions may be made, ones that can adapt to new urban challenges like population increase, climate change, and aging infrastructure.

By combining geographical data with digital twins, cities may create dynamic planning models that adapt to real-world situations and citizen needs. This allows for continuous and iterative procedures. Urban development becomes more robust, efficient, and sustainable, leading to better utilization of resources and a higher standard of living for all. Data interoperability, privacy issues, and governance frameworks are three crucial hurdles that must be overcome before Digital Twins can reach their full potential.

In the future, Digital Twins will become even more useful in creating cities that are prepared for the future because of advancements in AI, big data analytics, and immersive visualization. Innovation, inclusion, and environmental stewardship can be fostered through collaborative ecosystems encompassing communities, businesses, and governments through Digital Twin-enabled urban design. In the end, digital twins are a huge step toward making cities that are more intelligent, sensitive, and centered around their citizens so that they can survive and prosper in this complicated world.

## References

1. Jones D, Snider C, Nassehi A, Yon J, Hicks B Characterising the digital twin: A systematic literature review. [Accessed on 20 August 2025]
2. VanDerHorn E, Mahadevan S Digital twin: Generalization, characterization and implementation
3. Hexagon Digital Twin Report [accessed on 20 Aug 2025]
4. Kritzinger W, Karner M, Traar G, Henjes J, Sihn W (2018) Digital Twin in manufacturing: A categorical literature review and classification. IFAC-Papers OnLine 51(11):1016–1022
5. Gartner Research (2020) Emerging technologies: Digital twins drive business innovation
6. Zhou K, Fu C, Yang S (2016) Big data driven smart energy management. Renew Sustain Energy Rev 56:215–225
7. Virtual Singapore. https://tinyurl.com/295vvbkb [accessed on 25 Aug 2025]
8. Helsinki, Finland Case Study. https://tinyurl.com/5tthect2 [accessed on 10 Sep 2025]
9. Shanghai, China case study. https://tinyurl.com/3u3fhjz5 [accessed on 20 Sep 2025]
10. Amaravati Smart City Case Study. https://tinyurl.com/3wb42t5v [accessed on 20 Sep 2025]
11. World Bank (2021) The role of digital citizen engagement in modern governance. https://www.worldbank.org/en/topic/digitaldevelopment [accessed on 20 Aug 2025]
12. Accenture (2022) Metaverse and digital twin convergence. [Accessed on 20 Sep 2025]
13. UN-Habitat (2022) Smart cities and urban planning. https://unhabitat.org [accessed on 30 Sep 2025]

14. IEEE (2023) Virtual reality and augmented reality for urban design and planning. https://ieeexplore.ieee.org [accessed on 20 Sep 2025]
15. Sharma L, Garg PK (eds) (2025) Digital twins and simulation technology concepts and applications, 1st ed. Chapman and Hall/CRC. https://doi.org/10.1201/9781003582489
16. Sharma L, Garg PK (eds) (2024) Deep learning in internet of things for next generation healthcare, 1st ed. Chapman and Hall/CRC. https://doi.org/10.1201/9781003451846
17. Sharma L, Garg PK (eds) (2023) Technological prospects and social applications of society 5.0, 1st ed. Chapman and Hall/CRC. https://doi.org/10.1201/9781003324720
18. Sharma L, Carpenter M (eds) (2022) Computer vision and internet of things: Technologies and applications, 1st ed. Chapman and Hall/CRC. https://doi.org/10.1201/9781003244165
19. Sharma L, Garg PK (eds) (2021) Artificial intelligence: Technologies, applications, and challenges, 1st ed. Chapman and Hall/CRC. https://doi.org/10.1201/9781003140351
20. Sharma L (ed) (2020) Towards smart world: Homes to cities using internet of things, 1st ed. Chapman and Hall/CRC. https://doi.org/10.1201/9781003056751
21. Sharma L, Garg PK (eds) (2019) From visual surveillance to internet of things: Technology and applications, 1st ed. Chapman and Hall/CRC. https://doi.org/10.1201/9780429297922
22. Sharma L (2019) Object detection with background subtraction. LAP LAMBERT Academic Publishing, SIA OmniScriptum Publishing Brivibas gatve 197, Latvia, European Union. (ISBN: 978–613–7–34386–9)

# Chapter 6
# Digital Twins in Geospatial Science for Urban Planning and Development

**Yaman Hooda, Jimmy Mehta, and Lavanya Sharma**

**Abstract** The technological advancement of Artificial Intelligence is spanning in every direction with the invention of its sub-innovative technical developments. Digital Twins is such an application which allows to make a twin model of a physical system digitally. One of the most important advantages of Digital Twin is real-time Updation, which requires a robust system for collection, transmission and analysis of the data. Geospatial systems provide real-time information that blends collectively with the Digital Twin for making a data driven management system and thus, finds application in the domain of Urban Planning & Smart Infrastructure, Disaster Management, Environmental Monitoring, and Agriculture Industry. This chapter will deal with the amalgamation of the Digital Twin and Geospatial Sciences, towards developing a Smart and Sustainable Development. Moreover, it includes the discussion and frameworks of the emerging domains of Geospatial Artificial Intelligence (GeoAI) and Urban Digital Twin (UDT) Framework. The application of Digital Twin based GeoAI is twofold: contribution in the generation of spatial information and enhancing the efficiency of urban services. Urban Digital Twin (UDT) Framework is being employed in transforming the urban design practices with analyzation and optimizing the urban resource management systems, based on real-time data availability. The chapter also discusses the barriers and challenges towards the adaptation and implementation of Geospatial based Digital Twin Systems in planning, development, and management of urban smart infrastructure.

**Keywords** Digital twin · Geospatial technologies · Smart infrastructure · Urban infrastructure · Sustainable development · Smart built environment

---

Y. Hooda (✉)
Department Of Civil Engineering, Manav Rachna International Institute Of Research and Studies, Faridabad, Haryana, India
e-mail: yaman.set@mriu.edu.in

J. Mehta
Department Of Mechanical Engineering, Manav Rachna University, Faridabad, Haryana, India
e-mail: jimmymehta@mru.edu.in

L. Sharma
Department Of Information Technology, New Delhi Institute Of Management, Delhi, India

 
L. Sharma et al. (eds.), *Digital Twin Technology in Geospatial Science: Trends, Applications and Challenges*, Data-Intensive Research,
https://doi.org/10.1007/978-3-032-20311-3_6

## 1 Introduction

With the rapid expansion of applications of technological advancements such as Artificial Intelligence (AI), Internet of Things (IoT), and their sub-categories including Blockchain Technology, Digital Twin, Cloud Computing, and many more, along with the enhanced advancements in the domain of geospatial sciences, the approach of Smart Infrastructure is becoming popular (both in theory and practicality) in the domains of academics, politics, economics, and governmental relations [1, 2]. Urban planning and development play a crucial role in effective operations and implementations of different sectors of smart infrastructure by providing a vision, structured framework and coordination system for integrating various technologies towards built ecosystem in an efficient, sustainable and society-friendly way. The system of urban planning and development makes certain that the horizons of smart infrastructure, i.e., Energy Efficient Buildings, Intelligent Buildings, Intelligent Transportation System, Water and Sewage Management System and integration of digitalized technologies is integrated cohesively into the layout of a particular city without any occurrence of inefficiencies and fragmentations. The concept of "Smart Infrastructure" and "Intelligent Buildings" had been popularized and emphasized on the utilization of digital tools and data-driven mechanisms for enhancing the urban living standards. Among the digitalization landscape, the approach of "Digital Twin" is gradually being employed and recognized in the domain of urban planning and infrastructural management [3–6]. The concept of Digital Twins [DT] is based on the virtual complements of real-world elements, thus, bridging the physical & digital horizons of urban landscapes. This budding technology holds engaging applications towards urban development with real-time simulations and visualizations. The possible deployment of DT in urban development is immense, ranging from infrastructural planning to ecological sustainability by offering a holistic mechanism of data-driven procedure for addressing challenges such as traffic congestion and ecological deprivation [3, 7, 8]. Moreover, the integration of IoT and AI with DT improvise the process of predictive analytics, optimizing allocation of resources and helps in fostering of societal engagement [9]. While the possibilities of DT in urban planning are extensive, its application, especially concerning sustainability, remains in its infancy [3]. The amalgamation of dynamics of societal communities within the domain of DT is an emerging domain of research and innovation. Figure 1 depicts the possible framework architecture of application of DT in urban planning and development.

Geospatial sciences are paying a transformative and crucial role in urban development and planning by providing different technologies and tools, enabling improved analysis, simulation & visualization and decision-making mechanism. With the overgrowing trend of urbanization, change in climatic conditions, infrastructural stresses, and resource management, geospatial sciences are offering innovative approaches and solutions for an efficient-sustainable growth of the urban spaces [10]. Geospatial sciences cover the technological bandwidths of Remote Sensing, Geographic Information Systems [GIS], Global Positioning Systems [GPS], and Data Analytics [DA]. The above-mentioned tools of geospatial sciences are used for collection of the data,

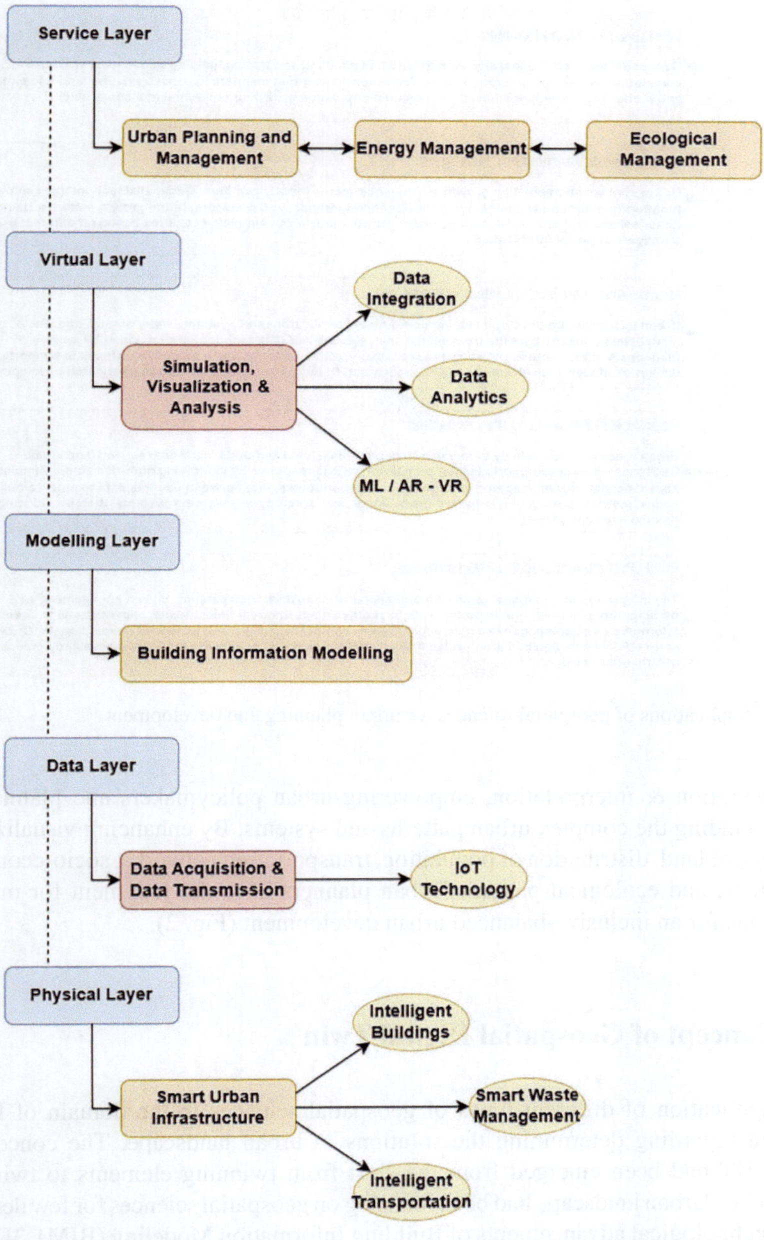


**Fig. 1** Architectural framework of digital twin towards smart urban planning and development

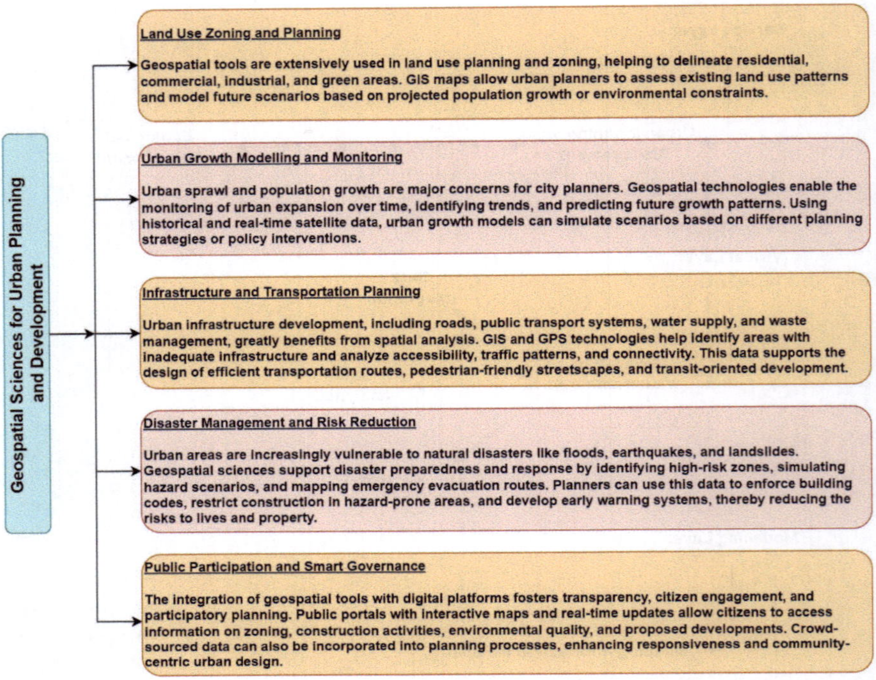


**Fig. 2** Applications of geospatial sciences for urban planning and development

its integration & interpretation, empowering urban policymakers and planners in understanding the complex urban patterns and systems. By enhancing visualization of usage of land, distribution of population, transportation networks, socio-economic dynamics, and ecological patterns, urban planners have the judgment for making decisions for an inclusive-balanced urban development (Fig. 2).

## 2 Concept of Geospatial Digital Twin

The application of different types of geospatial sciences in the domain of DT is explicit regarding determining the solutions of urban landscape. The concept of urban DT had been emerged from the shift from twinning elements to twinning cities [11]. Urban landscape had been focusing on geospatial sciences for few decades with technological advancements of Building Information Modeling (BIM), 3D City Models (3DCMs), and Spatial Data Infrastructures (SDIs) [12]. Many researchers had distinguished these technological services based on their characteristics, applications, and mechanisms for adaptation in urban development and planning [12]. These applications uses different temporal and spatial scales for interpretation of the data collected by different means of networking devices. Most of the studies also agreed

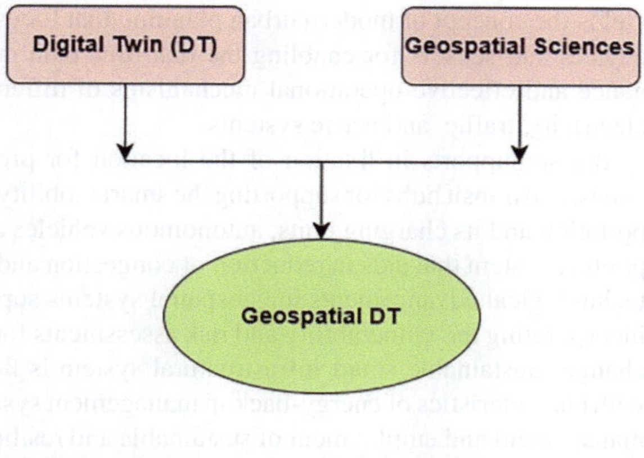


**Fig. 3** Geospatial DT: Geospatial sciences + Digital Twins (DTs)

on the fact that the functions of these applications can be enhanced by providing linkages with IoT & bi-directional real-time data integration. The reference of the geospatial sciences can be given in different ways in context to DT [13]. If "location" can be taken as a parameter, then the geographical location of the environment and the physical entity is to be considered a crucial data type, enabling the environment and entity to be pictured on a map [14] (Fig. 3).

The concept of "Geospatial Digital Twin" was first employed by Döllner [15] with regards to application of Machine Learning into 3D point clouding. Geospatial DT can be defined as "means for monitoring, visualizing, exploring, optimizing and predicting behavior and processes related to the corresponding physical entities". Other studies collectively summarized the concept of Geospatial DT as a digitalized replica of the spatial entities, where DL & ML methods are employed for interpreting, analysis, & organizing the data of 3D point clouding [16, 17]. For highlighting the data regarding any location through the application of DT, a term "Location-enabled DT" had been employed [14]. Geospatial DT is used for highlighting the fact that the domains of geospatial sciences are providing altitudinal framework of the data entities that are being presented in DT. Since all these terms had been utilized by different researchers based on their understanding for emphasizing and highlighting the relation between geospatial sciences and DT, there is still no clear distinction or characterization of domain of DTs that is based on the concepts and applicability of geospatial sciences.

With the amalgamation of geospatial technologies with the computation techniques, urban planning aids in prioritizing the infrastructural investments on societal needs, growth of the population and associated environmental impacts. The combined application helps in the setting up of renewable energy systems, placement of smart grid systems and IoT communication networks are useful for optimizing performance. The strategies of urban development ensured the fact that the designing and development of smart infrastructure must aligns with the objectives of sustainable development with the help of low-carbon transportation systems, green infrastructure, effective utilization of resources, and climate & disaster resilient infrastructural upgradation. The adaptation of computational techniques and geospatial sciences yields the concept of modern urban planning that focuses on the employment of IoT devices and sensors for enabling the real-time data monitoring, analytical maintenance and effective operational mechanisms of different utilities, including water, electricity, traffic, and waste systems.

It also supports in fixation of the location for providing roadways, pedestrian zones, and transit hubs for supporting the smart mobility options such as electric transportation and its charging units, autonomous vehicles and application-based vehicle pooling system that aids in reduction of congestion and pollution. The application of technological advancements in geospatial systems supports the urban developers in incorporating the vulnerability and risk assessments for natural disasters and climate change. Sustainable smart infrastructural system is flexible and adaptive in nature with characteristics of energy-backup management systems, sensor systems for flood management and employment of sustainable and resilient materials. One of the most important characteristics of the amalgamation of the technologies is providing the digital systems to the citizens for decision making; ensuring that smart infrastructure satisfies the actual needs with enhanced accessibility and thus, improving the quality of life of the society [18, 19].

## 3 GeoAI and Digital Twin for Urban Development

Smart urban planning and development thrives on the inclusion of digital and physical layers of infrastructure. Several models had been proposed that relies on digital framework of urban planning and development through technological advancements such as big data, digital twin, computational algorithm with Information and Communications Technologies [ICTs]. The emerging trend of digital twin for the past decades in the domain of Wireless Sensor Networks (WSNs), Earth Observation (EO), wireless technologies and Internet and Communications Technologies (ICTs) had been developed humongous amount of unstructured information at different stages of social, temporal, and spatial scales [20, 21]. The conjunction of the computational techniques of AI, DL, DT, and Data Sciences had enhanced the process of the collection of the data along with its processing, analysis, and interpretation. With the availability of enhanced computational infrastructural systems such as cloud computing platforms and Graphics Processing Units [GPU], the application of geospatial sciences with DT had been increased immensely [21].

Geospatial Artificial Intelligence [GeoAI] has been emerged as an interdisciplinary analytical spatial domain that combines the fundamentals of GIS with technological foregoing revolutions. It is the technological solution for computation of intensive geospatial scenarios & heralds a fresh domain of data-intensive investigation, supplementing the theoretical, empirical, & computational models [21, 22]. The concept of GeoAI depends on two different procedural mechanisms: Top-Down [knowledge-driven] and Bottom-Up [Data-driven]. With considering the theme of the study, the latter approach is useful as it is considered to be an inductive procedure; that heavily relies on the applications of DT techniques and thus, is becoming a conventional mechanism in GeoAI in conjunction with ML; due to their characteristics of identification and solution of hidden patterns for making effective prediction [23, 24]. The intersection of DT and GeoAI is creating a powerful feedback mechanism loop between the real physical world its digital replica. The approach of GeoAI is enhancing the functionality and intelligence of DT by providing systematic capabilities for detecting anomalies, interpretation of spatial data patterns, and prediction of future scenarios. Simultaneously, DTs are providing structured and rich data environment for effective operation of GeoAI algorithms. In urban planning, a digitalized twin of a city may integrate the real-time information from the employed building energy systems, traffic sensors, and climatic conditions. The algorithms of GeoAI can analyze the temporal-spatial information for identification of congestion hotspots, predicting transportation requirements for future and assessing the influence of fresh development projects. This will permit the urban planners in simulation of different planning scenarios in DT & optimizing the process of decision-making based on AI-driven perceptions. The applications of GeoAI-DT integration for urban planning and development is summarized below.

### *3.1 Smart Cities*

GeoAI permits DTs of the cities for learning the patterns in usage of energy, mobility, quality of air and management of waste. Urban planners can simulate the extent of modification in the urban designing and policy-making with overall efficiency of the system and improving the quality of life.

### *3.2 Environmental Monitoring*

The combination of GeoAI with DT for natural environmental conditions allows constant monitoring of deforestation activities, levels of water under the surface of earth and overall health of the ecosystem. Predictive modeling that had been trained on historical spatial data aids the urban planners and developers in forecasting changes and support conservation efforts (Fig. 4).

### *3.3 Disaster Management and Risk Reduction*

A digitalized twin of an area can incorporates the real-time sensor and satellite data, whereas the GeoAI can forecast the nature and post-consequences of any disaster such as floods, wildfires, or disease outbreaks. With the conjunction of the two technical advancements, the strategies for the emergency response can be virtually examined before being employed in reality.

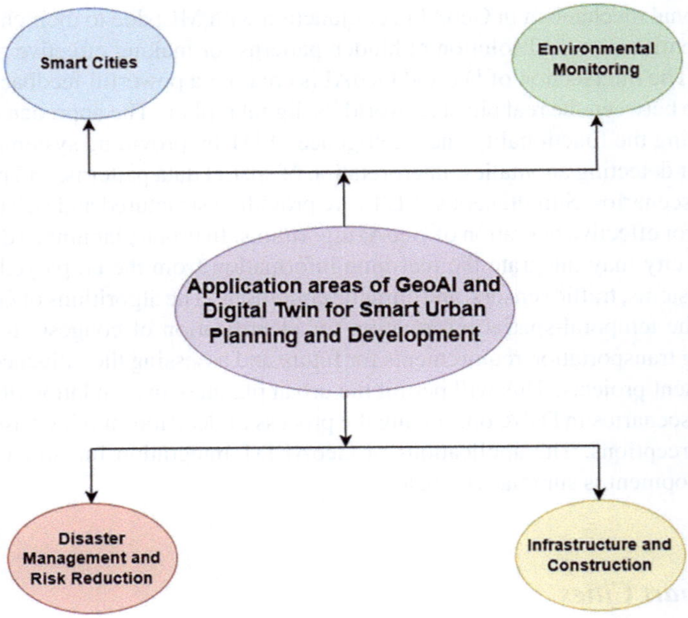


**Fig. 4** Application areas of GeoAI and digital twin for smart urban planning and development

### *3.4 Infrastructure and Construction*

In the domain of asset management and construction industry, DTs of infrastructure and buildings are being integrated with GeoAI prototypes for monitoring structural health, optimizing resource consumptions and improving lifecycle management processes.

In spite of extreme potential, the integration of DT and GeoAI faces challenges and barriers in its application, such as managing data privacy, ensuring data interoperability, building scalable AI models and maintaining high-quality geospatial datasets. Furthermore, the computational demand of simulation and analysis of real–time data may prove to be intensive. As such, GeoAI could serve as a potential solution framework to address the multifaceted geospatial problems facing smart urban development. Indeed, the complexity of smart-city problems calls for a hybrid approach that not only integrates both inductive and deductive methodologies, but also informs and is informed by theory. Nevertheless, with the advancements in the domain of 5G networking, edge computing, and cloud computing, the capability of deployment of GeoAI within DT is speedily enlightening. Alos, interoperable platforms and open data standards are helping in seamless fusion of digital modeling and geospatial sciences.

## 4 Urban Digital Twin Framework

Urban Digital Twin Framework, commonly known as UDT Framework, can be defined as a virtual representation of urban ecosystem dynamically; integrating real-time information & improvised modeling methods for simulation, analyzation, prediction, and optimization of urban ecosystems. UDT is serving as a stage for providing complex models; facilitating data-driven decision making for smart sustainable urban planning and development. Supplementing with UDT, the applications of AI incorporate a wide spectrum of complex models & algorithms that permits computer systems for emulating cognitive procedures (problem-solving, reasoning, and learning). Predictive analysis is crucial in proactive design and planning, permitting cities in adaptation regarding the growing challenges, identification of bottlenecks, and enhancing the urban efficiency. In conjunction with geospatial sciences, UDT models enable the prediction of urban drifts, development and assessment of strategies regrading planning and development, and identification of prime solutions providing an inclusive groundwork for sustainable & adaptive urban ecosystem [25–27]. Geospatial sciences are used for providing foundation geo-information to the UDT system. While geospatial sciences are used in providing and analysis of static spatial data such as topography, zoning, and infrastructural networks, UDT framework is based on the integration of the collected real-time and historical information for constant feedback & simulating process. In effect, geospatial sciences focus on "where" & "what," while UDT reactions include "what if" & "what next." The main advantage lies in enhancing urban dynamics, strengthening of urban resilience, empowering designers, planners & stakeholders in making evidence-based decisions for promoting the concept and vision of sustainable smart built environment. Some of the domains that is directly affected by the amalgamation of UDT framework with geospatial sciences towards urban planning and development are

- Allocation of resources,
- Infrastructural development,
- Optimization of transportation facilities,
- Air quality,
- Energy management, and
- Conservation of biodiversity.

The prime objective of complex computational model of UDT-geospatial sciences is to provide a dynamic & comprehensive visualization or urban ecosystem including natural resources, infrastructure, social dynamics, and natural resources. With the help of simulating behavior and interactions with urban horizons, the integrated system allows the stakeholders in analyzing, visualization and optimizing urban ecosystems for making efficient decisions and experiments various conditions for urban sustainable development [25, 28, 29, 37, 45, 46]. The application domains of geospatial sciences based UDT framework for urban planning and development are summarized (Fig. 5).

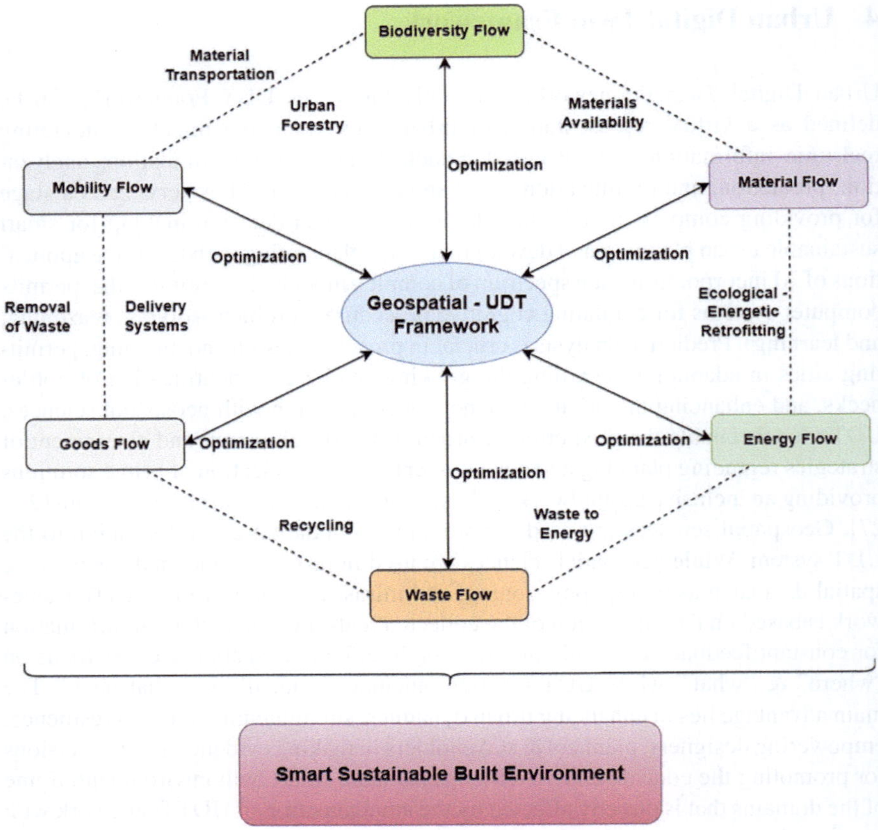


**Fig. 5** Geospatial-UDT framework for smart sustainable built environment

## *4.1 Land Use Planning*

The different techniques of geospatial sciences provide spatial understandings regarding zoning regulations, availability of land and density of population. UDTs simulates forthcoming circumstances such as change in pattern of land use, growing trend of population and demands of houses, aiding stakeholders in understanding long-term impressions and making efficient decisions on land management.

## *4.2 Infrastructural Planning and Management*

UDT framework integrates GIS-based patterns of buildings, roads, power grids, and pipelines with the help of real-time information from sensors. The advantage lies in identification of faulty devices or mechanisms, predicting failure junctions and methods, and optimization of maintenance schedules. Geospatial sciences support long-term infrastructural planning and UDTs permits the real-time procedures & experimentation of modifications before their physical execution.

### 4.3 *Environmental Sustainability and Monitoring*

The layers of GIS such as water bodies, vegetation cover, and sources of pollution helps in analyzing urban ecological conditions. The simulation from the UDT Framework aids in the development of urban ecosystem with changing patterns of the air quality and pollution levels towards "Green" urban planning and development.

### 4.4 *Traffic Management and Smart Mobility*

The technology of geospatial sciences aids in visualization of density of traffic, public transport routes, & hotspots for accidents. UDTs simulates the flow of the traffic by utilizing real-time data from traffic lights & vehicles helping the urban planners in testing the changes for signal timings, road expansions, reduction in traffic congestion and improvising mobility services.

### 4.5 *Societal Engagement and Governance*

GIS/GPS dashboards & interactive mapping services updates the public regarding prevailing and upcoming urban projects. UDT improves this application by empowering citizens to explore the developments virtually, welcome feedbacks and visualizing the impact of the changes in the societal neighborhood.

## 5 Challenges and Barriers

The integration of technological advancements of geospatial sciences and DT showcased wide variety of applications for urban development and planning due to their characteristics of collection, visualization, and simulation of real-time data; along with impressive decision-making potential. Nevertheless, the adaptation of this unconventional technical framework faces some barriers and challenges technically, economically, socially, and organizationally [47, 48].

The most primary challenge is the quality, availability, and integration of the data. DT system requires high-resolution and comprehensive geospatial information regarding infrastructure, land use, traffic, environment, and socio-economic aspects [self]. For developing nations, this information is either outdated, unavailable, unstandardized, or fragmented. Integration of the data from different sources & confirming interoperability among services and systems is a key barrier, especially for the cases where the departments of the government tail the variation in data formats and protocols. Infrastructural limitations and technological complexities are also impeded in the adaptation of Geospatial-Digital Twin frameworks. For building up such technical infrastructure, the system requires a robust computational framework, cloud storage, real-time sensor networks & highly equipped modeling tools. Most of the urban development and planning organizations lack access to these advanced technological systems, which are becoming essential for updating the models and maintaining the update with real-time information. Furthermore, 3D HD models and platforms for real-time simulation require skilled personnel and powerful computing resources for its operation and maintenance [30].

Another critical challenge is the unavailability of capacity building & technical experts within the administrators, urban planners & developers, and other related stakeholders such as architects and engineers. A successful implementation of Geospatial-Digital Twin systems necessitates inter-disciplinary knowledge of urban systems engineering, geoinformatics, artificial intelligence, and data sciences. The gap in the technical awareness and training leads to inability or reluctance towards their adoption. Economical constraints also play a significant role in restricting the deployment of Geospatial-Digital Twin system. The development and maintenance of DT of a particular area is capital-intensive [32, 39, 40]. The finances involved in establishment of sensor networks, attaining high-quality information, and investment in hardware infrastructure and related software are significant. Most of the urban localities and municipalities are operating under fitted budgets and thus, finds it difficult in sustaining or justifying such huge investments without any public–private partnerships or external fundings. Regulatory and institutional challenges are complicating the conditions as the amalgamation of the DT technology with geospatial science requires cross-sectoral association and sharing of data between private and public bodies. However, rigid governance hierarchy, unclear data governance policies and lack of coordination frequently generate bottlenecks. Problems related to the privacy and security of the data and their ownership are also emerging, especially when dealing with real-time or sensitive private information from the society [32–34, 43, 44]. Social trust and acceptance are significant deliberations. Society can be sceptical about data collection & surveillance events connected with real-time DT applications [35, 36, 40–42]. Apprehensions regarding the method of collection, storage, and usage of data can be taken into consideration with ethical guidelines, transparent communication, and robust information defensive guidelines and policies.

# 6 Conclusion

The prime objective of this chapter is to enlighten the individual and combined role of Geospatial Sciences and Digital Twins for the planning and development of urban infrastructure. DT is playing a crucial role in the domain of smart urban development and planning by creation of virtual, real-time imitations of physical ecosystem, integrating the information from IoT sensors & devices, AI and sources of geospatial sciences such as Remote Sensing and GIS for simulation, monitoring and predicting the urban ecosystems. The integrated technological framework of DT and geospatial sciences are proving to be the boon for smart urban development and planning by optimizing infrastructural systems, improving traffic movements, enhancing energy efficiency & evaluating ecological impacts. The correlation between DT and GeoAI is transformative and symbiotic. GeoAI supplements the layer of predicative capabilities and spatial intelligence to DT, whereas, DT offers a structured, rich context for GeoAI for generation of actionable information. The synergy between geospatial sciences and UDT framework characterizes a paradigm shift in urban infrastructural planning & development. Together, these two technological advancements empower cities to grow sustainably with smarter planning and faster response system, meeting the needs of both current & future generations.

**Acknowledgements** Thank you for your cooperation and contribution.

# References

1. Banerjee T (2014) General introduction. In: Banerjee T (ed) Urban design: Critical concepts in urban studies, vol 1, pp 1–24. Routledge, London, UK. The Idea of Urban Design
2. Batty M (2013) The new science of cities. MIT Press
3. Mylonas G, Kalogeras A, Kalogeras G, Anagnostopoulos C, Alexakos C, Muñoz L (2021) Digital twins from smart manufacturing to smart cities: A survey. IEEE Access: Pract Innov, Open Solut 9:143222–143249
4. Depretre A, Jacquinod F, Mielniczek A (2022) Exploring digital twin adaptation to the urban environment: Comparison with cim to avoid silo-based approaches. ISPRS Ann Photogramm, Remote Sens Spat Inf Sci
5. Deren L, Wenbo Y, Zhenfeng S (2021) Smart city based on digital twins. Comput Urban Sci 1:1–11
6. Elsehrawy R, Kumar B, Watson R (2021) A digital twin uses classification system for urban planning & city infrastructure management. J Inf Technol Constr 26:832–862
7. Dembski F, Wössner U, Letzgus M, Ruddat M, Yamu C (2020) Urban digital twins for smart cities and citizens: The case study of Herrenberg, Germany [Article]. Sustainability (Switzerland), 12(6):2307. https://doi.org/10.3390/su12062307
8. Erol T, Mendi AF, Doğan D (2020) Digital transformation revolution with digital twin technology. In: 2020 4th international symposium on multidisciplinary studies and innovative technologies (ISMSIT)
9. Fattahi Tabasi S, Rafizadeh HR, Andaji Garmaroudi A, Banihashemi S (2023) Optimizing urban layouts through computational generative design: Density distribution and shape optimization. Arch Eng Des Manag

10. Hooda Y (2022) IoT and remote sensing. In: Computer vision and internet of things. Chapman and Hall/CRC, pp 111–140
11. Lei B, Stouffs R, Biljecki F (2023) Assessing and benchmarking 3D city models. Int J Geogr Inf Sci 37:788–809
12. Jeddoub I, Nys GA, Hajji R, Billen R (2023) Digital Twins for cities: Analyzing the gap between concepts and current implementa tions with a specific focus on data integration. Int J Appl Earth Obs Geoinf 122:103440
13. Jones D, Snider C, Nassehi A, Yon J, Hicks B (2020) Characterising the digital twin: A systematic literature review. CIRP J Manuf Sci Technol 29:36–52
14. Ellul C, Stoter J, Bucher B (2022) Location-enabled digital twins–understanding the role of NMCAS in a European context. Isprs Ann Photogramm Remote Sens Spat Inf Sci X-4/W2:53–60
15. Döllner J (2020) Geospatial artificial intelligence: potentials of machine learning for 3D point clouds and geospatial digital twins. PFGJ Photogramm Remote Sens Geoinf Sci 88:15–24
16. Rathore MM, Shah SA, Shukla D, Bentafat E, Bakiras S (2021) The role of AI, machine learning, and big data in digital twinning: Asystematic literature review, challenges, and opportunities. IEEE Access 9:32030–32052
17. Hooda Y, Kuhar P, Sharma K, Verma NK (2021) Emerging applications of artificial intelligence in structural engineering and construction industry. In: Journal of physics: Conference series, vol 1950, no 1. IOP Publishing, p 012062
18. Batty M (2018) Digital twins. Environ Plan B, Urban Anal City Sci 45(5):817–820 (Web)
19. Batty M, Axhausen KW, Giannotti F et al (2012) Smart cities of the future. Eur Phys J-Spec Top 214:481–518
20. Goodchild MF (2019) Geography and GIScience: An evolving relationship. The Canadian Geogr 63(4):530–539. https://doi.org/10.1111/cag.12554
21. Li W (2020) GeoAI: Where machine learning and big data converge in GIScience. J Spat Inf Sci 20:71–77. https://doi.org/10.5311/JOSIS.2020.20.658
22. Gahegan M (2020) Fourth paradigm GIScience? Prospects for automated discovery and explanation from data. Int J Geogr Inf Sci 34(1):1–21. https://doi.org/10.1080/13658816.2019.1652304
23. Janowicz K, Gao S, Grant MK, Hu Y, Bhaduri B (2020) GeoAI: Spatially explicit artificial intelligence techniques for geographic knowledge discovery and beyond. Int J Geogr Inf Sci 34(4):625–636. https://doi.org/10.1080/13658816.2019.1684500
24. Razavi S (2021) Deep learning, explained: Fundamentals, explainability, and bridgeability to process-based modelling. Environ Model & Softw 144, Artic 105159. https://doi.org/10.1016/j.envsoft.2021.105159
25. Salunke AA (2023) Reinforcement learning empowered digital twins: pioneering smart cities towards optimal urban dynamics. EPRA Int J Res Develop (IJRD). https://doi.org/10.36713/epra13959
26. Gkontzis AF, Kotsiantis S, Feretzakis G, Verykios VS (2024) Enhancing urban resilience: Smart city data analyses, forecasts, and digital twin techniques at the neighborhood level. Futur Internet 16(2):47. https://doi.org/10.3390/fi16020047
27. Ye X, Du J, Han Y, Newman G, Retchless D, Zou L, Zhang L, Cai Z (2023) Developing human-centered urban digital twins for community infrastruc ture resilience: A research agenda. J Plann Lit 38(2):187–199. https://doi.org/10.1177/08854122231167319
28. Bibri SE, Huang J, Jagatheesaperumal SK, Krogstie J (2024) The synergistic interplay of artificial intelligence and digital twin in environmentally plan ning sustainable smart cities: a comprehensive systematic review. Environ Sci Ecotechnol 20:100433. https://doi.org/10.1016/j.ese.2024.100433
29. Wu J, Wang X, Dang Y, Lv Z (2022) Digital twins and artificial intelligence in transportation infrastructure: classification, application, and future research directions. Comput Electr Eng 101 (September 2021):107983. https://doi.org/10.1016/j.compeleceng.2022.107983
30. Akter S, McCarthy G, Sajib S, Michael K, Dwivedi YK, D'Ambra J, et al (2021) Algorithmic bias in data-driven innovation in the age of AI

31. Mutambik I, Almuqrin A, Alharbi F, Abusharhah M (2023) How to encourage public engagement in smart city development—learning from Saudi Arabia. Land 12(10):1851
32. Braun T, Fung BC, Iqbal F, Shah B (2018) Security and privacy challenges in smart cities. Sustain Cities Soc 39:499–507
33. Cui L, Xie G, Qu Y, Gao L, Yang Y (2018) Security and privacy in smart cities: Challenges and opportunities. IEEE Access 6:46134–46145
34. Mehta J, Hooda Y, Aggarwal Y (2025) 13 emergingdigital twin application technology of toward development of smart infrastructure. Digit Twins Simul Technol: Concepts Appl 186
35. Wang H, Chen X, Jia F, Cheng X (2023) Digital twin-supported smart city: Status, challenges and future research directions. Expert Syst Appl 217, Article 119531
36. Weil C, Bibri SE, Longchamp R, Golay F, Alahi A (2023) Urban digital twin challenges: A systematic review and perspectives for sustainable smart cities. Sustain Cities Soc 99, Article 104862
37. Sharma L, Garg PK (eds) (2025) Digital twins and simulation technology concepts and applications, 1st ed. Chapman and Hall/CRC. https://doi.org/10.1201/9781003582489
38. Sharma L, Garg PK (eds) (2024). Deep learning in internet of things for next generation healthcare, 1st ed. Chapman and Hall/CRC. https://doi.org/10.1201/9781003451846
39. Sharma L, Garg PK (eds) (2023) Technological prospects and social applications of society 5.0, 1st ed. Chapman and Hall/CRC. https://doi.org/10.1201/9781003324720
40. Sharma L, Carpenter M (eds) (2022) Computer vision and internet of things: Technologies and applications, 1st ed. Chapman and Hall/CRC. https://doi.org/10.1201/9781003244165
41. Sharma L, Garg PK (eds) (2021) Artificial intelligence: Technologies, applications, and challenges, 1st ed. Chapman and Hall/CRC. https://doi.org/10.1201/9781003140351
42. Sharma L (ed) (2020) Towards smart world: Homes to cities using internet of things, 1st ed. Chapman and Hall/CRC. https://doi.org/10.1201/9781003056751
43. Sharma L, Garg PK (eds) (2019) From visual surveillance to internet of things: Technology and applications, 1st ed. Chapman and Hall/CRC. https://doi.org/10.1201/9780429297922
44. Sharma L (2019) Object detection with background subtraction. LAP LAMBERT Academic Publishing, SIA OmniScriptum Publishing Brivibas gatve 197, Latvia, European Union. (ISBN: 978–613–7–34386–9)
45. Tonmoy MR, Hossain MM, Dey N, Mridha MF (2025) MobilePlantViT: A Mobile-Friendly Hybrid ViT for generalized plant disease image classification. arXiv preprint arXiv:2503.16628
46. Dey N, Bhatt C, Ashour AS (2018) Big data for remote sensing: Visualization, analysis and interpretation. Springer, Cham, p 104
47. Dey N, Wagh S, Mahalle PN, Pathan MS (eds) (2019) Applied machine learning for smart data analysis. CRC Press
48. Solanki VK, Katiyar S, BhashkarSemwal V, Dewan P, Venkatasen M, Dey N (2016) Advanced automated module for smart and secure city. Procedia Comput Sci 78:367–374

# Chapter 7
# GIS-Based Digital Twins for Environmental Monitoring and Climate Change

**Miloš Rusnák, Goga Tomáš, Michaleje Lukáš, and Hamid Afzali**

**Abstract** Development in new technology, big data processing and artificial intelligence creates new possibilities for environmental assessment and analysis, where a virtual model of the ecosystem can simulate and provide evidence of ecosystem health and anthropogenic impact. Digital twins represent a robust approach where predictions and scenarios have also led to an understanding of deeper interconnections and chain processes, and how this vulnerable system responds to human intervention and climate change. Linking to real-time monitoring can increase the response to negative impacts and is applied to disaster risk reduction through improved recovery, policy-making, resilience, forecasting, resource allocation, scenario testing, lessons learned, community engagement, and decision-making. Here, three different examples are presented where a precise 3D model is connected with real-time data and the decision-making process for solving problems in smart cities and urban greenery and forestry, flood disaster management by digital twins, and application of digital twins of the river to detect ecosystem degradation and management.

**Keywords** Digital twin · GIS · Environmental assessment · Climate change

M. Rusnák (✉) · M. Lukáš · H. Afzali
Department of Physical Geography, Geomorphology and Natural Hazards, Institute of Geography Slovak Academy of Sciences, Bratislava, Slovakia
e-mail: geogmilo@savba.sk

M. Lukáš
e-mail: geoglumi@savba.sk

H. Afzali
e-mail: geogafza@savba.sk

G. Tomáš
Department of Geoinformatics, Institute of Geography, Slovak Academy of Sciences, Bratislava, Slovakia
e-mail: geogleho@savba.sk

H. Afzali
Department of Physical Geography and Geoecology, Comenius University in Bratislava, Mlynská dolina, Bratislava, Slovakia



L. Sharma et al. (eds.), *Digital Twin Technology in Geospatial Science: Trends, Applications and Challenges*, Data-Intensive Research,
https://doi.org/10.1007/978-3-032-20311-3_7

## 1 Introduction

A geographic information system (GIS) is a technical system that focuses on the spatial information storage and plays a key role in data collection, planning, environmental protection, decision support, resource protection, sustainable land use or natural disaster monitoring [29]. Spatial data and the development of GIS technology are rooted in cartography and field mapping by traditional geodetic survey and drawing techniques. Throughout the 1960s, computer advances led to the development of GIS as a basic spatial tool for storing information. Advances in technology development bring new web interactive platforms, cloud storage and computing, and complex management of digital data. Finally, a new generation of virtual geographic environment evolves to a generation of geospatial digital twins that integrate GIS, big data, AI support tools, scenario simulation, and a robust decision-making platform [61].

In parallel with the development of technology, anthropogenic modification and human impact on Earth are increasing rapidly. To understand environmental processes, the functioning of ecosystems, and monitoring, the concept of digital twins is an empowering tool for exploring the consequences of human impact and has been widely applied for decision management [14]. Digital twins create a precise model of real natural objects in a virtual environment that represents the state of nature and is interconnected with other real-time or near real-time datasets (digital data) that can model, predict, and visualise natural processes. Cloud computing and big data development connected with artificial intelligence enable a complex system understanding of the physical environment.

For instance, digital twins and satellite real-time situation data from optical and radar sensors are used to predict flood extent and emergency response plans for the management of hazards [10, 27]. A precise 3D model enables detailed modelling and is attractive to visualise problems with the active participation of citizens, or recognises critical infrastructure [40, 65]. In emergency management, digital twins enable an iterative process of adaptation planning and dynamic decision-making. With the increasing impact of extreme weather events, heavy rainfalls and prolonged drought periods, water and flood protection management are key for reducing the negative impact on both human presence in urban areas and also to mitigate biodiversity loss due to higher anthropogenic pressure resulting from human adaptation to climate change. Here, the digital twin concept is essential to modelling and predicting real situations to reduce the consequences of flood events, but involves stakeholders and the local community to find a solution that uses nature-based and green solutions to enhance ecology and ecosystem quality. Moreover, a robust system can repeatedly predict adaptation measures with changing initial conditions. Durden [7] presented specific use cases for using digital twins for environmental management, such as habitat identification, status of the benthic community, identification of species, invasive species, optimisation of surveys, conservation measurement, testing climate scenarios or communication with the public.

In the urban areas, smart city concepts are developed for efficient mapping of the urban environment, where 3D datasets are connected with a spatial GIS layer of architecture, urban networks, green infrastructure or internal government databases, such as the cadastre or socio-economic distribution of inhabitants [61]. Adding future scenario prediction and stakeholders' insight can transform the smart city infrastructure into a digital twin concept [27]. Loss of greenery, increase in urban population, deterioration of air quality, increase in urbanisation, temperature and the phenomenon of the urban heat island [17, 64] are challenges that are mutually interactive and have a multiplicative effect on urban areas. Sustainable development (industrial, commercial, or agricultural) necessarily requires assessing environmental impact and evaluating ecosystem health [7], which is in accordance with the UN Sustainable Development Goals.

Here, the application of digital twin within the concept of ecological management, restoration and environmental monitoring is presented. Nature and ecosystems are complex systems with mutual interaction of biotic and abiotic factors, strongly affected by human interference or direct and indirect influence. The digital twin model, generated from novel and precise 3D data, connected with remote sensing, cloud processing, field measurement, and human pressure with stakeholders' requirements, is crucial for bridging the gaps in the scientific, technological and management capabilities for a deep understanding of the system and its future scenario prediction.

## 2 Technology for 3D Model Generation

Digital twin is characterised as a digital representation of a physical component, a digital replica of physical assets, processes, or systems [19]. In the spatial domain and environmental assessment, digital twins are the concept of the interactive representation of the Earth, or parts of the surface of the Earth [14]. This concept was originally developed for the industry by NASA to simulate real-time behaviour for the Apollo 13 spacecraft. The real-time correlation between digital twins and their physical characterisation is fundamental to understanding and enabling climate change adaptation and environmental management.

Various Earth simulation systems emerged during the last several years as the European Commission's Destination Earth [16], NASA's Earth System Digital Twin, or NVIDIA's Earth [14]. Basically, digital twins are built on Earth observation data and supported by machine learning approaches and big data processing. Spatial data are collected by non-contact methods with different types of sensors. Multiple data integrations are available based on accurate GNSS positioning to generate a seamless model of the environment with different spatial scales and a system with sub-systems integration. Geospatial application of digital twins is based on integration of virtual geographic environment (VGE) and digital twin, pointed to real-time integrations and actual Earth state from digital models [62]. Based on the process of data obtaining, we can divide spatial sources that can be used for the digital twin of the Earth, created for office-based sources, field data, drone imaging, aerial imaging and data from space orbits (Fig. 1).

The first type of data is vectorisation and processing of available secondary and existing data sources in a GIS or CAD environment, where original 2D data in local or geographical coordinates are transformed to 3D datasets, adding a vertical dimension directly from datasets or when the original data only contains XY information extracted and combined from another data source or is estimated [29]. In these cases, old or historical plans, maps, cartographic data and technical drawings are used to extract topography from contour lines, building footprints with elevation, structure of the objects and spatial position of electric lines, wires, and specific spatial information. These data sources combine both spatial information (XYZ coordinates) and specific thematic information (power lines, building construction material, building wall material, forest type, temporal changes, etc.).

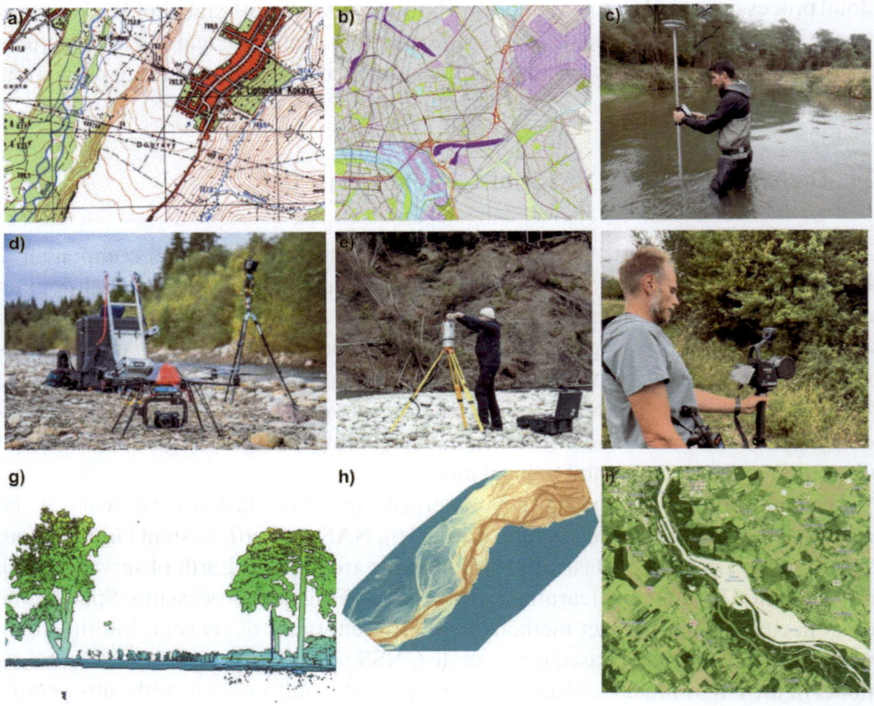

**Fig. 1** Geospatial sources for creation digital twins and 3D virtual datasets: topographic maps (**a**), city plans (**b**), GNSS measurement in the field (**c**), drone mapping with UAS and camera (**d**), terrestrial laser scanning (TLS) measurement, handled SLAM Lidar FARO Orbis (**f**), point cloud of urban greenery from SLAM lidar (**g**), airborne elevation model from ground point cloud (**h**), and satellite images from SENTINEL 2 in NDVI band visualisation (**i**)

Field or ground measurements are used for local and large-scale analyses where traditional methods, such as geodetic measurement (levelling, tachymetry or GNSS technology), can be used to map 3D objects with millimetre precision. Modern technology of scanning that has emerged over the last decades can be used for rapid data collection and automatic 3D model generation of the area of interest. This method is built on a photogrammetric approach with optical passive sensors (photographs) or active sensors emitting a laser beam to scan the object structure. Optical remote sensing uses a traditional photogrammetric approach to detect 3D objects and feature extraction by identifying the same points in multiple images, with the camera's position and orientation, by applying triangulation to detect 3D position. New development in computer visualisation techniques using the SfM (Structure from Motion) algorithm, which was developed in the 1970s [58] and later used in photogrammetry as SfM-MVS—Structure-From-Motion Multi-View Stereo Photogrammetry [6, 9, 13]. Similar to classical photogrammetry, the algorithm uses image (photograph) overlap to identify identical objects from multiple angles, automatically calculating internal and external orientation elements without entering predefined and targeted control points [34, 46]. An active sensor called LiDAR (light detection and ranging) sends and receives laser signals to retrieve elevation with exact positioning calculated based on sensor position and trajectory [57], or used SLAM (simultaneous localisation and mapping) based on concepts in computational geometry and computer vision with simultaneous tracking of an unknown environment while simultaneously keeping track on the previous scanned model by inner precise IMU unit [8, 26, 31, 32, 43]. Basically, ground field lidar measurement can use a TLS (terrestrial laser scanning) or MLS (mobile laser scanning) approach to survey single objects or small areas of land with high speed and measurement accuracy and precision [37]. During mobile scanning, we can use a handheld scanner or a sensor that can be carried by a moving platform on a car or a boat. Additionally, the lidar survey is now supported by commercial mobile phones, such as iPhone Pro, which can generate a colorised dense point cloud of the real world [26]. Both of the methods, which use SfM or Lidar technology, are usually combined with a drone survey to scan larger areas by using either a flying or a boat platform [31]. The last decade of development has approved drones—UAS (unmanned aerial systems)—as mobile and effective mapping systems for obtaining the earth surface spatial information and virtual digital twins [46, 47, 52].

Aerial data that can be partially included, as well as data from UAS, are usually considered as data captured by airborne scanning, including a range of sensors, mainly metric cameras, lidar sensors, or radar. Spatially, airborne scanning data cover large areas and are usually available in the form of national datasets. Aerial images can be used for the generation of photogrammetric models or for the colourisation of point clouds. The density of the point cloud generated from higher altitudes is lower, but it can sufficiently cover huge areas.

Remote sensing provided from space is now most popular for regular data acquisition and an automatic data processing chain. One of the most popular satellite programs is LANDSAT, where the first satellite constellation was launched in 1972 and covers the whole Earth. Satellite sensors capture information about the Earth's surface and atmosphere by measuring the reflected energy. These systems are categorised as active and passive sensors, such as ground field measurement. Passive sensors detect visible, near-infrared, and short-wave infrared spectra of the solar radiation energy reflected from Earth's ground. Active sensors are usually radar (Synthetic Aperture Radar Sensors) or Lidar, which emit their own energy signal and capture a reflected signal with surface characteristics. From SAR altimetry or LiDAR scanning, it is possible to directly detect the elevation and the vertical component of the object's spatial dimension. Nowadays, Earth observation satellites are human eyes tracking changes on the Earth and are widely used as a primary source for monitoring and assessment of demonstrating the effects of global warming and climate change. Similarly, satellite Earth observation is used to detect sea ice coverage shrinking, sea-level rise, glacial movements, earthquakes, forest fires, landslides, extreme rainfall events, predict changes in land change-induced desertification, loss of tropical forests, illegal deforestation, evaporation, transpiration, and soil moisture. Furthermore, they are key elements for monitoring the Earth's carbon cycles, water quality, greenhouse gases, the ozone layer, and the prediction of extreme weather events, and providing early warning for disasters.

## 3 System Design of Environmental Monitoring Based on Digital Twins

Digital twins in environmental monitoring and assessment of the effect of climate change realistically represent complex natural and urban systems in the form of a 3D model and with an effective system of functioning natural and ecosystem processes, linked with real-time data and offering an informed decision-making solution related to system state definition, system prediction and changes based on real time data input, disaster modelling, prediction, post-disaster recovery and data driven stakeholder incorporation [53, 66]. The system is rooted in a combination of realistic 3D spatial datasets with geospatial real-time analyses or simulation, which enable synchronisation, bidirectional communication, and real-time control interaction across the systems [11]. The system is built on information flow within the framework of Earth coordinates or presented 3D models and consists of: (i) a tangible space of physical objects, (ii) a virtual space of digital objects,and (iii) 3 data flow between information exchange [11].

The concept of using digital twins for environmental monitoring relies on the development of effective systems that depend on enabling new technologies (Fig. 2). Basically, the creation of virtual models and the definition of physical relationships for the real functioning of the system are necessary for effective virtual reality that reflects the natural environment in virtual space. Connecting geospatial datasets for 3D model generation based on different technologies such as point clouds, rasters or vector formats, with pre-defined coordinates, with real-time monitoring and social or economic data defines a basic database for knowledge, a process-oriented basement for data-driven intervention and decision-making process. Interconnections and system integration depend on user aims and defined user application targets. For instance, an inaccessible ecosystem in the middle of a rain forest will be studied with a focus on ecological diversity, sustainable development and application of space remote sensing data instead of detailed economic and social data. Furthermore, digital twins of the city park will be more connected with stakeholder participation, economic development or city plans, and perceptions of inhabitants living in the proximity of the 3D objects. The definition and complexity of digital twins reflect the application's targets, where models are built on static 3D visualisations or fully functional systems capable of adaptive, bidirectional operations [7].

Human intervention reflected in biodiversity loss or disaster risk reduction is connected with forecasting and prediction, incorporating climate change changes with different scenarios for future planning and management [14]. Climate data can

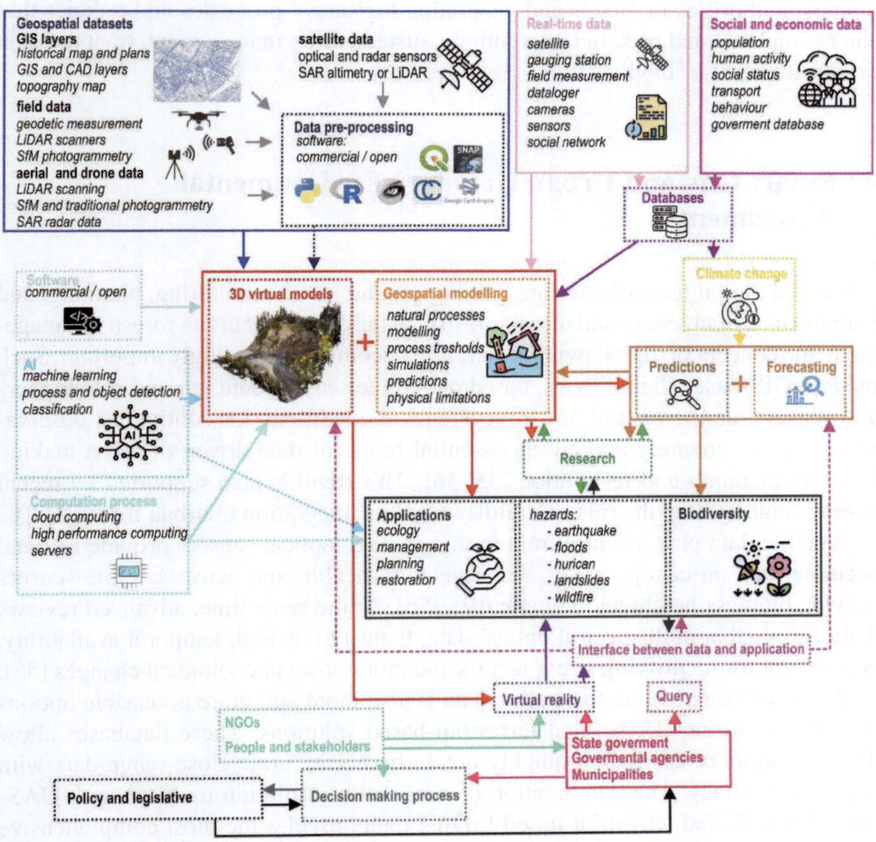


**Fig. 2** Conceptual design for using digital twins for environmental monitoring and assessment

be entered into geospatial modelling and can be used for data analysis or process visualisation in virtual reality (VR). Virtual reality throughout digital twins is used to involve the public and stakeholders in the decision process, learning and understanding the ecological process and for future actions and decisions. Creation of standardised protocols and frameworks is essential to facilitate seamless data integration, interoperability (data formats, communication protocols, and application programming interfaces (APIs)) and standards for geospatial data play a key role in this process of compatibility [20]. Nevertheless, the main challenges, such as data processing, high-performance computing, computing and storage space, deep understanding of processes, and data standards, are limitations for using real digital twin models for wide use in environmental analyses [7].

User interaction (government, governmental agencies or municipalities) can be preserved by different formats and platforms (web, mobile, apps, virtual reality, augmented reality, API). The whole system can also be effective by providing critical data for decision-making processes, enhancing collaboration among stakeholders, authorities or NGOs and generating legislative processes and policies that can be implemented back in interventions, sustainability management, revitalisation or improvement of biodiversity.

## 4 Smart City and Urban Greenery Environmental Assessment

Advanced digital technologies are gaining ground in the monitoring, planning, and maintenance of greenery and urban forestry. In the context of urban forestry management, the concept of digital twins is therefore becoming increasingly important. Such modern GIS-based digital twins, based on remote sensing data, ground data loggers and sensors, and terrestrial 3D surveys, create conditions for continuous observation of urban greenery and are an essential basis for data-driven decision-making in urban planning and restoration [55, 56]. This trend is also supported by recent reviews highlighting the role of multisource data integration in urban forestry [30].

Satellite data play a central role in this context–optical sensors provide fast and accurate data on canopy cover and vegetation health, and active satellite sensors provide biomass height and volume data [56]. At the same time, advanced reviews demonstrate that multispectral optical data, thanks to its high temporal availability, is excellent for identifying stress factors and monitoring phenological changes [55].

A crucial complement to satellite data is also more and more accessible options in airborne, drone (UAS), and terrestrial-based solutions. These databases allow their operators to easily and quickly obtain highly accurate close-range data with geodetic accuracy. The combination of centimetre-resolution multispectral, UAS-based LiDAR, and terrestrial-based LiDAR data provides the most comprehensive (and most massive) 3D data layers for urban forestry monitoring at the individual tree level. Such 3D approaches have been emphasised for canopy cover detection and tree health assessment [30].

Climate change in the 21st century has also highlighted the challenges of monitoring urban heat islands. Integrating real-time data into automated monitoring and management of urban greenery is highly desirable in this area [38, 48]. The integration of ground sensors (temperature, humidity, or soil moisture) with thermal remote sensing sensors (satellite, airborne, and UAS-based) can help identify local heat islands with fine granularity [64]. Latest studies further confirm the role of satellite monitoring in detecting UHI hotspots and guiding targeted planting [63]. Hotspots identified in this way will enable urban planners to identify better areas that require targeted planting of urban greenery (and trees) to cool these hotspots. Integrating real-time data will enable the creation of automated plans for intelligent irrigation systems. Furthermore, linking local microclimate data with canopy cover data supports the creation of strategies for opening shaded structures or scheduling misting systems in public green spaces on extremely hot days.

Accurate 3D data on greenery is the basis for creating a digital twin for smart cities focused on monitoring urban parks and green areas. 3D models at the tree level created using LiDAR and photogrammetry methods provide sufficiently accurate information about the structure of tree crowns and canopy cover [55]. Additionally, using terrestrial-based laser scanning, we can supplement this data with equally important information about trunk diameter and tree branching under the canopy [30]. Integrating this data in a broader context with other GIS-based urban databases allows urban planners to improve the interaction between the built-up environment and urban greenery (e.g., with engineering networks, transportation, and buildings).

Lastly, an important feature of the smart city concept integrated into digital twins is predictive modelling, scenario planning, and simulation capabilities. Digital twins enable the design of future urban green development in various scenarios, e.g., how planting a row of taller trees might reduce wind speeds along a street; how a new building could cast a shadow on nearby vegetation; how canopy cover will develop in various stress situations. These simulation approaches are increasingly used to forecast effects on temperature, air quality, runoff, and biodiversity [55]. Individual models can evaluate the outcomes of different strategies: what if a city plants more trees in a specific district, or conversely, what if development removes a portion of green space? The models can forecast effects on temperature, air quality, runoff, and biodiversity. Moreover, they can be used to assess risks, such as the impact of extreme events on urban trees (wind, extreme drought), and can serve as a basis for long-term strategies or resource/budget allocation for future planting.

A key indicator in urban green space management is the identification of ecosystem services provided by urban forestry. Indicators obtained within the smart-cities concept allow for the quantification of ecosystem services and monitoring of their performance over time, e.g., the amount of carbon captured in urban forestry, the volume of solid particles filtered from the atmosphere, and the amount of rainwater captured [56]. The most important ecosystem service that city residents immediately notice is the cooling effect of trees. Other monitorable ecosystem services include providing habitats for birds and pollinators, noise reduction, and others [30].

All technological and environmental components—integrated into a coherent urban forestry management strategy—in line with the smart cities' initiative, are very effective in gaining a comprehensive view of urban forests and creating a comprehensive digital platform for data integration. Integrating data from satellites, drones, terrestrial devices, and real-time data loggers supports more efficient urban greenery planning and monitoring. The smart-cities concept integrated into digital twins enables the balancing of socio-environmental inequalities and the identification of tree diseases, droughts, and even vandalism. These aspects are strongly related to equitable governance frameworks [28], urban green equity dimensions [36], and more recent concepts of equitable forest governance [42]. Appropriately set policies enable the automation of urban forestry restoration processes by directing interventions where needed to restore the required ecosystem services as quickly as possible. The policies and approaches based on data-driven decisions must transcend administrative and time barriers to ensure continuity in green infrastructure management even during political changes in the local government.

## 5 Flood Disaster Management by Digital Twins

Flood hazard is part of hydrological natural hazards, which are defined as any natural process or phenomenon that may cause loss of life, injury or other health impacts, property damage, loss of livelihoods and services, social and economic disruption or environmental damage [59]. Hydrological events were the most common natural disasters in the world. Due to climate change and socioeconomic changes, it is expected that in 2080, the European Union's annual affected population will increase from 200,000 to 360,000 and annual damage from 5.5€ billion to 53€ billion if measures are adapted with consideration of future changes of climate change [45]. In flood risk management, the first and most crucial step is flood risk assessment [1]. Thus, knowledge of the spatial distribution of risk and its magnitude is the basis for decision-making [35]. Information provided by flood risk assessment helps decision-makers to tailor the best strategies and choose the most suitable measures to decrease adverse effects of floods on population, economy and environment. Flood risk management adopts the principle of complexity, which takes into account the whole basin and its physico-geographical components, such as geology, topography, hydrology, climatology, land cover and socio-economic components, as well as the spatial distribution of population and its economic activities and their state of vulnerability to flooding.

The concept of digital twins provides the necessary tools that help fulfil the potential of the principle of complexity. In urbanised areas, it helps with managing drainage networks by monitoring and detecting real-time water levels, predictive maintenance and managing urban drainage systems [25]. Using datalogger data and physically based models helps detect flooding and monitor blockage, and the need for maintenance of stormwater networks in urban environments [2]. Collecting data with the help of citizen participation (citizen science) through open-source platforms provides locations of flooded areas that can help to validate and calibrate models of flooding [40].

Modern methods and technologies provide an opportunity to study flood hazard in greater detail. Remote sensing is one of these tools. High frequency of satellite passing coupled with a variety of sensors provides near-real time data about (i) changes in land cover and its properties such as composition, roughness and density of vegetation, (ii) soil moisture and vegetation health, (iii) precipitation estimates and precipitation rates, (iv) evaporation and transpiration. Satellite data with the help of AI can help us detect properties of different classes of land cover, such as the absence or presence of snow cover, foliage during summer and winter periods and management of meadows, which directly affect the generation of high runoffs and floods. With the rapid passage of the satellite, this can provide near-real-time input data for rainfall-runoff modelling.

Using UAS and LiDAR technology, we can create a detailed DEM as an input for semi-automatic detection of the channel and extraction of basic hydraulic parameters, roughness of the channel and identification of obstacles in it [33]. These data also provide fundamental information for connectivity between the channel and the riparian zone, which is directly impacted by floods. Climatological and hydrological data are the background for rainfall-runoff modelling and hydraulic modelling for the extent of flood for different scenarios of seasons during the year. Stakeholders and citizens are a crucial part of this process, because they provide valuable information from the catchment, as they have the best knowledge of the affected area and its properties. Farmers provide data about agricultural practices and crops. Disperse crops like corn with smaller roughness to create conditions for higher runoff on exposed slopes around urbanised areas. Uncut grass can be swept by heavy rain and create a tilled roof effect that can prevent water from infiltrating the ground.

## 6 Digital Twins of the River: Example of Ecosystem Degradation and Management

River systems rank among the most biologically diverse and productive ecosystems on Earth, strongly influenced by river flow-related physical processes, such as erosion, sediment deposition, periodic flooding, and groundwater-surface water interactions. Naturally, they exhibit significant spatial and temporal variability, fostering a wide array of species, habitats, and ecological processes. Understanding these processes is crucial for maintaining ecologically sound conditions that support dynamic aquatic ecosystems. The last decades have seen an increased focus on river quality, driven by the implementation of the Water Framework Directive (2000/60/EU) to improve river health. Currently, the EU's biodiversity strategy for 2030 (The European Green Deal, Action Environment and Oceans) emphasises restoring the degraded ecosystems with an impact on river transformation into good ecological status. Human influence on European watercourses is substantial, often leading to irreversible environmental changes. Anthropogenic modification, grade-control structures and channelisation resulted in channel narrowing, sediment transport transformation and incision in many rivers in Europe [3, 5, 12, 18, 21, 44, 54, 60]. Indirect impact is presented by flow condition changes due to outer impact into the basin, as well as land cover changes.

The impact of climate change creates complex cause-and-effect chains related to altered environmental conditions and stressors in the riverine landscape. These changes are primarily associated with fluctuations in precipitation and temperature. River channels, which are extensively utilised for water resources, are particularly vulnerable to global climate change stressors. This vulnerability is heightened due to their relative isolation from terrestrial landscapes [41]. Feedback between flow, sediment dynamics, channel forms, and riparian vegetation is explained as part of the function and changes of the riparian ecosystem in space and time. The fluvial geomorphic processes reciprocally interact and control the impact of riparian vegetation [39].

The digital twin's approach is revolutionising the detailed monitoring and assessment of river ecosystems. The monitoring of riparian ecosystems is essential for understanding the way that systems respond to stressors and management interventions. Intensive field sampling can provide useful insights into the status and trends of local systems. However, this approach can be labour-intensive and costly due to its dynamic nature, the large area monitored, and the relative inaccessibility of riparian ecosystems [23, 49]. Digital twins provide a powerful tool for monitoring riparian zones over long durations and large areas. Advances in satellite and airborne sensor technology, improved spectral and scanning capabilities (such as radar and lidar), and innovative field mapping tools (including echo sounding, drones, and terrestrial laser scanners) have greatly enhanced our ability to study these dynamic interactions within riparian ecosystems in detail connected with real-time monitoring based on satellite sensors, dataloggers, gauging stations and information from local people and tourist by social media or voluntary mapping using the apps on mobile phones (e.g., AMBER project and barrier atlas: https://portal.amber.international/). Point clouds serve as valuable resources for topographic analyses and the identification of channel transformation [4]. In recent years, the advent of low-cost monitoring through unmanned aerial vehicles (UASs) and the application of SfM photogrammetry [22] have opened new ways for detecting high-resolution topographic, hydraulic and vegetation changes in river systems.

The main challenge in digital twins' models is capturing the different parameters that shape the riparian ecosystem. This includes understanding the physical factors and morphology of the channel, the flow parameters (velocity, depth and temperature), the riparian vegetation, sediment movement and the way individuals and populations mutually interconnect with processes [46]. Some clear advantages of GIS, remote sensing and digital twins for assessing freshwater biophysical properties are the cost, the product accuracy, the data continuity, and the availability of programming software or personnel skills [15]. The landform structure, substrate grain size and stratigraphy, geochemical properties, and water availability create the basic framework for riparian plant communities. A real virtual model of the ecosystem offers efficient monitoring and detection of these entities.

## 6.1 *Study Area and Digital Twin of the River*

Digital twin of the last multichannel braided-wandering river in Slovakia (Belá River) was generated by a combination of field measurement, drone survey and airborne Lidar scanning (Fig. 3). The overall length of the river is 23.6 km and originates from the Western and High Tatra Mountains. With its riparian zone, the river system creates a supra-regional importance biocorridor and is included in the NATURA 2000 network. During the 20th Century, the river system underwent significant transformation due to climate change, human direct and indirect impacts, exhibits clear signs of channel incision, and is fully embanked through urbanised areas [24, 47].

A digital twin model was generated from aerial LiDAR in the form of an LAZ point cloud reaching a vertical accuracy of 0.03 m and an average density for the last reflective points of 18 points/m$^2$. Airborne Lidar was combined with the SfM-UAS-generated point cloud from 2015 and 2022, and TLS measurements between 2016 and 2020 (12 scan horizons). The output from PhotoScan included orthophotomosaics with a resolution of 5 cm and a point cloud with a density ranging between 200 and 400 points per m$^2$. The geospatial dataset was connected with field measurements (cross-sections, sediment analyses, geophysical survey, vegetation census) and real-time data or near real-time data, such as hydraulic data from the gauging station and satellite data from Sentinel-2 and Planet Scope, using Google Earth Engine (GEE) cloud computing and PyQGIS library for monitoring and responses of the river ecosystem to fluctuation in climate conditions, direct or indirect human impact.

## 6.2 *Digital Twin: A Tool for Effective River Monitoring*

Digital twin model pointed to significant channel transformation, narrowing, morphological degradation and incision. Human intervention is presented in the study area by transforming the natural river channel into a channelised section and land cover transformation of upper catchment areas (afforestation resulted in trapping of sediments). Channel incision (Fig. 4a) was detected from repeated point cloud models and pointed to the extreme channel incision of up to 4.05 m in the degraded section from 1945, and SfM-UAS survey and cross-sections confirm these detected ongoing incisions up to 1.39 m between 2007 and 2015. A significant deficit of sediment after the emptying of the channel infill with headcut propagation led to the loss of biodiversity, decline of groundwater level and rare species in the Natura 2000 protected area. Overall, from the point cloud, erosion of 31,876 m$^3$ of sediment was detected, and the total net amount was -22,759 m$^3$. In this situation, only active management aimed at increasing sediment flow in the upper part of the river basin can mitigate the

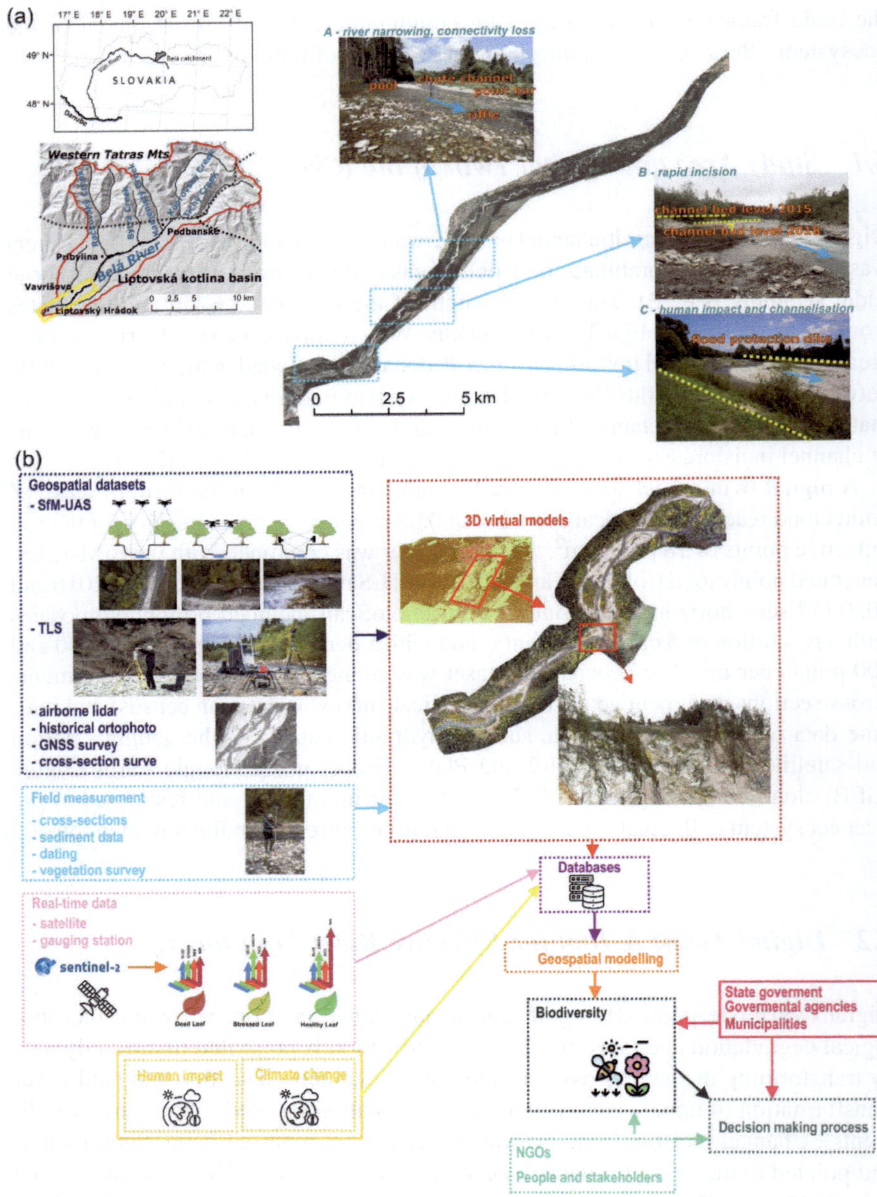


**Fig. 3** Localisation of the Belá River system in Slovakia with the general catchment situation with tributaries and orthophoto of the riparian zone (**a**). Examples of the 3 most important negative transformations of the river channel are presented in photographs A, B and C. Processes of digital twin generation, connecting with satellite and hydraulic data for biodiversity and ecological assessment, are presented in (**b**)

intensity of river degradation. However, these activities contrast with the flood control measures announced in the lower part of the Belá River. Morphological evolution and vegetation state were monitored based on vegetation and moisture indices calculated from Sentinel-2 satellite data. Vegetation structure reflects active river dynamics, human forestry management, and climate change disturbances related to bark beetle outbreaks (Fig. 4b). The upper catchment of the Belá River was affected by forest damage caused by an exceptional windstorm in 2004, and this event triggered a serious (the largest ever recorded in Slovakia) bark beetle outbreak, which ended in forest clearing and vegetation rejuvenation. The spatial distribution of vegetation and moisture indices was found to be independent of the distance from the channel, with notable differences observed across different channel platforms. Seasonal variations in vegetation and moisture indices followed the local hydrological conditions and the annual phenology curve. Real-time or near real-time monitoring is secured from Sentinel-2 images, which result from NDVI analyses pointed to negative reactions of different types of vegetation to morphological changes in degraded river sections (Fig. 4c).

## 7 Conclusion

The digital twin framework presented a robust approach for environmental assessment and analysis. Current state and development of new technologies and scanning approaches enable scanning of a vulnerable ecosystem and creating a virtual model. For this model, geospatial modelling tools and simulation can be applied to understand processes and impacts. Also, predictions and scenarios led to an understanding of deeper mutual interconnections and chain-related processes and how this vulnerable system reacts to human interventions and climate variations. Connection with real-time monitoring can increase reaction to negative consequences and is applied for disaster risk reduction by enhancing recovery, policy-making, resilience, prediction, resource allocation, scenario testing, lessons learned, community engagement, and decision-making. It is still necessary and important to strengthen knowledge transfer between the technicians, scientists, and managers or decision-makers to exploit the digital twin concept more effectively. In the present day, it is improving due to the technical availability of user-friendly methods and their routine application in research and virtual reality. The question is the effectiveness of remote sensing, point clouds and large spatial datasets techniques for information mining and transfer of pixel data to process-oriented information and the integration of quantitative and qualitative information. Implementation of near real-time monitoring is complicated, expensive and requires network technology. Moreover, questions related to open data repositories and policies with knowledge and know-how sharing can increase the green and ecological application of new technologies.

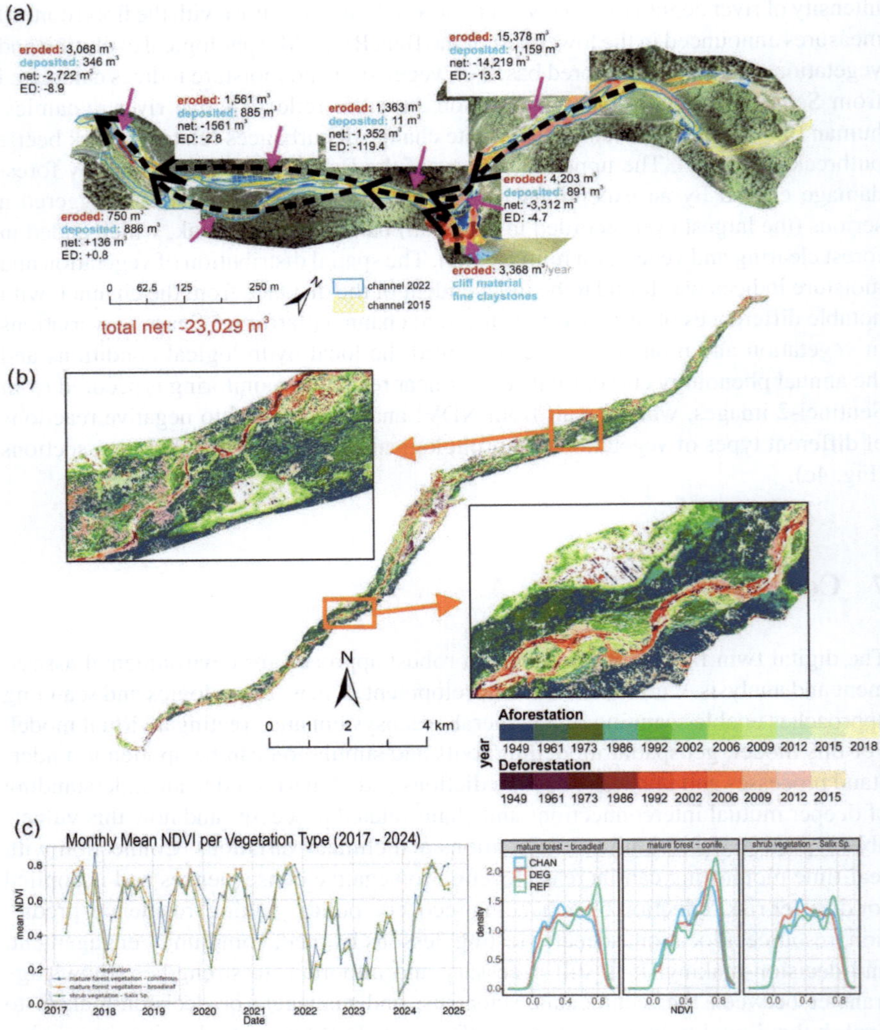


**Fig. 4** Transformation of the river channel with rapid incision up to 1.39 m detected from SfM-UAS point clouds (**a**), vegetation transformation in the riparian zone with afforestation and deforestation dating from historical aerial images (**b**) and reaction of different riparian vegetation types to channel morphological changes detected based on NDVI calculated from satellite data (**c**)

**Acknowledgement** *The study was supported by the Science Grant Agency (VEGA) of the Ministry of Education, Science, Research and Sport of the Slovak Republic and the Slovak Academy of Sciences [Grant No. 02/0016/24] and by the Slovak Research and Development Agency under Contract No. APVV-23-0265. This work is part of the research within the COST Action INTUF: European Network for the Integrative Approach of Urban Forestry (CA23148). The authors are grateful to the Geodesy, Cartography and Cadastre Authority of the Slovak Republic (ÚGKK SR)* *for providing LiDAR data. The historical aerial images were provided by the Topographic Institute in Banská Bystrica, and the orthophotos by Eurosense Slovakia, s.r.o. and the Geodetic and Cartographic Institute Bratislava (GKÚ) with the National Forest Centre (NLC).*

## References

1. Atkinson M, Bouchon S, Heikkila AM, Nordvik JP (2006) Report on user requirements supporting risk and vulnerability management of systemic and industrial risks. http://www.crh.noaa.gov/unr/?n=1972_Floo
2. Bartos M, Kerkez B (2021) Pipedream: An interactive digital twin model for natural and urban drainage systems. Environ Model & Softw 144:105120. https://doi.org/10.1016/J.ENVSOFT.2021.105120
3. Bertalan L, Rodrigo-Comino J, Surian N, Šulc Michalková M, Kovács Z, Szabó S, Szabó G, Hooke J (2019) Detailed assessment of spatial and temporal variations in river channel changes and meander evolution as a preliminary work for effective floodplain management. The example of Sajó River, Hungary. J Environ Manag, 248:109277. https://doi.org/10.1016/j.jenvman.2019.109277
4. Bertoldi W, Gurnell AM, Drake NA (2011) The topographic signature of vegetation development along a braided river: Results of a combined analysis of airborne lidar, color air photographs, and ground measurements. Water Resour Res 47(6). https://doi.org/10.1029/2010WR010319
5. Brenna A, Marchi L, Borga M, Zaramella M, Surian N (2023) What drives major channel widening in mountain rivers during floods? The role of debris floods during a high-magnitude event. Geomorphology:108650. https://doi.org/10.1016/J.GEOMORPH.2023.108650
6. Dietrich JT (2016) Riverscape mapping with helicopter-based structure-from-motion photogrammetry. Geomorphology 252:144–157. https://doi.org/10.1016/j.geomorph.2015.05.008
7. Durden JM (2025) Environmental management using a digital twin. Environ Sci & Policy 164:104018. https://doi.org/10.1016/J.ENVSCI.2025.104018
8. Dušeková L, Herich P, Pukanská K, Bartoš K, Kseňak Ľ, Šveda J, Fehér J (2024) Comparison of non-contact measurement technologies applied on the underground glacier—the choice for long-term monitoring of ice changes in Dobšiná ice cave. Remote Sensing, 16(20). https://doi.org/10.3390/rs16203870
9. Fonstad MA, Dietrich JT, Courville BC, Jensen JL, Carbonneau PE (2013) Topographic structure from motion: a new development in photogrammetric measurement. Earth Surf Process Landf 38(4):421–430. https://doi.org/10.1002/esp.3366
10. Ge C, Qin S (2025) Urban flooding digital twin system framework. Syst Sci Control Eng 13(1). https://doi.org/10.1080/21642583.2025.2460432
11. Grieves M (2015) Digital twin: Manufacturing excellence through virtual factory replication. https://www.researchgate.net/publication/275211047_Digital_Twin_Manufacturing_Excellence_through_Virtual_Factory_Replication
12. Hajdukiewicz H, Wyżga B, Zawiejska J (2017) Twentieth-century hydromorphological degradation of Polish Carpathian rivers. Quat Int. https://doi.org/10.1016/j.quaint.2017.12.011
13. Harwin S, Lucieer A (2012) Assessing the accuracy of georeferenced point clouds produced via multi-view stereopsis from Unmanned Aerial Vehicle (UAV) Imagery. Remote Sensing 4(6):1573–1599. https://doi.org/10.3390/rs4061573
14. Hazeleger W, Aerts JPM, Bauer P, Bierkens MFP, Camps-Valls G, Dekker MM, Doblas-Reyes FJ, Eyring V, Finkenauer C, Grundner A, Hachinger S, Hall DM, Hartmann T, Iglesias-Suarez F, Janssens M, Jones ER, Kölling T, Lees M, Lhermitte S, Vossepoel FC et al (2024) Digital twins of the Earth with and for humans. Commun Earth & Environ 5(1):1–11. https://doi.org/10.1038/s43247-024-01626-x

15. Hestir EL, Brando VE, Bresciani M, Giardino C, Matta E, Villa P, Dekker AG (2015) Measuring freshwater aquatic ecosystems: The need for a hyperspectral global mapping satellite mission. Remote Sens Environ 167:181–195. https://doi.org/10.1016/j.rse.2015.05.023
16. Hoffmann J, Bauer P, Sandu I, Wedi N, Geenen T, Thiemert D (2023) Destination earth—a digital twin in support of climate services. Climate Services 30:100394. https://doi.org/10.1016/j.cliser.2023.100394
17. Holec J, Feranec J, Šťastný P, Szatmári D, Kopecká M, Garaj M (2020) Evolution and assessment of urban heat island between the years 1998 and 2016: case study of the cities Bratislava and Trnava in western Slovakia. Theor Appl Climatol 141(3–4):979–997. https://doi.org/10.1007/s00704-020-03197-1
18. Holušová A, Poledniková Z, Vaverka L, Galia T (2023) Spatiotemporal dynamics and present perception of gravel bars in natural and regulated environments. Sci Total Environ 892:164711. https://doi.org/10.1016/J.SCITOTENV.2023.164711
19. Huang J, Wu P, Li W, Zhang J, Xu Y (2024) Exploring the applications of digital twin technology in enhancing sustainability in civil engineering: A review. In: SDHM structural durability and health monitoring, vol 18, Issue 5, pp 577–598. Tech Science Press. https://doi.org/10.32604/sdhm.2024.050338
20. Igbokwe E, Ezeh F, Oliha A (2024) Twin technology in surveying and geoinformatics. Int J Multi-Discipl Res Growth Eval 5(3):520–526. www.allmultidisciplinaryjournal.com
21. James LA (1997) Channel incision on the Lower American River, California, from streamflow gage records. Water Resour Res 33(3):485–490. https://doi.org/10.1029/96WR03685
22. James MR, Robson S (2014) Mitigating systematic error in topographic models derived from UAV and ground-based image networks. Earth Surf Process Landf 39(10):1413–1420. https://doi.org/10.1002/esp.3609
23. Sharma L, Garg PK (eds) (2019) From visual surveillance to internet of things: Technology and applications, 1st ed. Chapman and Hall/CRC. https://doi.org/10.1201/9780429297922
24. Kidová A, Radecki-Pawlik A, Rusnák M, Plesiński K (2021) Hydromorphological evaluation of the river training impact on a multi-thread river system (Belá River, Carpathians, Slovakia). Scientific Reports 11(1):6289. https://doi.org/10.1038/s41598-021-85805-2
25. Kim Y, Oh J, Bartos M (2025) Stormwater digital twin with online quality control detects urban flood hazards under uncertainty. Sustain Cities Soc 118:105982. https://doi.org/10.1016/J.SCS.2024.105982
26. Kovanič Ľ, Peťovský P, Topitzer B, Blišťan P, Tokarčík O (2025) Analysis of the qualitative parameters of mobile laser scanning for the creation of cartographic works and 3d models for digital twins of urban areas. Applied Sciences 15(4):2073. https://doi.org/10.3390/app15042073
27. Lagap U, Ghaffarian S (2024) Digital post-disaster risk management twinning: A review and improved conceptual framework. Int J Disaster Risk Reduct 110:104629. https://doi.org/10.1016/J.IJDRR.2024.104629
28. Lawrence A, De Vreese R, Johnston M, Konijnendijk van den Bosch CC, Sanesi G (2013) Urban forest governance: Towards a framework for comparing approaches. Urban For & Urban Green 12(4):464–473. https://doi.org/10.1016/J.UFUG.2013.05.002
29. Li X , Yue J, Wang S, Luo Y, Su C, Zhou J, Xu D, Lu H, Martínez-Graña M (2023). Development of geographic information system architecture feature analysis and evolution trend research. Sustainability 16(1):137. https://doi.org/10.3390/SU16010137
30. Li X, Chen WY, Sanesi G, Lafortezza R (2019) Remote sensing in urban forestry: Recent applications and future directions. In: Remote sensing, vol 11, Issue 10. MDPI AG. https://doi.org/10.3390/rs11101144
31. Liao J, Zhou J, Yang W (2021) Comparing LiDAR and SfM digital surface models for three land cover types. Open Geosciences 13(1):497–504. https://doi.org/10.1515/geo-2020-0257
32. Marshall A, Morrison RR, Jones B, Triantafillou S, Wohl E (2024) Handheld lidar as a tool for characterizing wood-rich river corridors. River Res Appl. https://doi.org/10.1002/rra.4239
33. Sharma L (ed) (2020) Towards smart world: Homes to cities using internet of things, 1st ed. Chapman and Hall/CRC. https://doi.org/10.1201/9781003056751

34. Micheletti N, Chandler JH, Lane SN (2015) Structure from motion (SFM) photogrammetry. In: Geomorphological techniques: Vols. XL-5/W4, pp 37–44. https://doi.org/10.5194/isprsarchives-XL-5-W4-37-2015
35. Mudashiru RB, Sabtu N, Abustan I, Balogun W (2021) Flood hazard mapping methods: A review. J Hydrol 603:126846. https://doi.org/10.1016/J.JHYDROL.2021.126846
36. Nesbitt L, Meitner MJ, Sheppard SRJ, Girling C (2018) The dimensions of urban green equity: A framework for analysis. Urban For & Urban Green 34:240–248. https://doi.org/10.1016/J.UFUG.2018.07.009
37. Nováková M, Gallay M, Šupinský J, Ferré E, Asti R, de Saint Blanquat M, Bajolet F, Sorriaux P (2022) Correcting laser scanning intensity recorded in a cave environment for high-resolution lithological mapping: A case study of the Gouffre Georges, France. Remote Sens Environ 280:113210. https://doi.org/10.1016/J.RSE.2022.113210
38. Sharma L, Garg PK (eds) (2021) Artificial intelligence: Technologies, applications, and challenges, 1st ed. Chapman and Hall/CRC. https://doi.org/10.1201/9781003140351
39. Osterkamp WR, Hupp CR (2010) Fluvial processes and vegetation—glimpses of the past, the present, and perhaps the future. Geomorphology 116(3–4):274–285. https://doi.org/10.1016/j.geomorph.2009.11.018
40. Park S, Kim J, Kim Y, Kang J (2024) Participatory framework for urban pluvial flood modeling in the digital twin era. Sustain Cities Soc 108:105496. https://doi.org/10.1016/J.SCS.2024.105496
41. Perkins DM, Reiss J, Yvon-Durocher G, Woodward G (2010) Global change and food webs in running waters. Hydrobiologia 657(1):181–198. https://doi.org/10.1007/S10750-009-0080-7/FIGURES/3
42. Pike K, Nesbitt L, Conway T, Day SD, Konijnendijk C (2024) What is equitable urban forest governance? A systematic literature review. Environ Sci & Policy 162:103951. https://doi.org/10.1016/J.ENVSCI.2024.103951
43. Pukanská K, Bartoš K, Gašinec J, Pašteka R, Zahorec P, Papčo J, Bella P, Andrássy E, Dušeková L, Bobíková D, Kseňak Ľ (2024) Methodological approaches to survey complex ice cave environments—the case of Dobšiná (Slovakia). Front Environ Sci 12. https://doi.org/10.3389/fenvs.2024.1484169
44. Rinaldi M, Wyżga B, Surian N (2005) Sediment mining in alluvial channels: physical effects and management perspectives. River Res Appl 21(7):805–828. https://doi.org/10.1002/rra.884
45. Rojas R, Feyen L, Watkiss P (2013) Climate change and river floods in the European Union: Socio-economic consequences and the costs and benefits of adaptation. Glob Environ Chang 23(6):1737–1751. https://doi.org/10.1016/J.GLOENVCHA.2013.08.006
46. Rusnák M, Goga T, Michaleje L, Michalková MŠ, Máčka Z, Bertalan L, Kidová A (2022) Remote sensing of Riparian ecosystems. Remote Sensing 14(11):2645. https://doi.org/10.3390/rs14112645
47. Rusnák M, Kaňuk J, Kidová A, Lehotský M, Piégay H, Sládek J, Michaleje L (2024) Inferring channel incision in gravel-bed rivers: Integrating LiDAR data, historical aerial photographs and drone-based SfM topo-bathymetry. Earth Surf Process Landf. https://doi.org/10.1002/esp.5840
48. Sharma L, Carpenter M (eds) (2022) Computer vision and internet of things: Technologies and applications, 1st ed. Chapman and Hall/CRC. https://doi.org/10.1201/9781003244165
49. Sharma L (2019) Object detection with background subtraction. LAP LAMBERT Academic Publishing, SIA OmniScriptum Publishing Brivibas gatve 197, Latvia, European Union, FEBRUARY 2019. (ISBN: 978-613-7-34386-9)
50. Rusnák M, Sládek J, Kidová A (2018a) Využitie UAV technológie pre klasifikáciu a mapovanie krajiny vo fluviálnej geomorfológii. Geogr Časopis-Geogr J 70(2):141–160. https://doi.org/10.31577/geogrcas.2018.70.2.08
51. Rusnák M, Sládek J, Kidová A, Lehotský M (2018b) Template for high-resolution river landscape mapping using UAV technology. Measurement 115:139–151. https://doi.org/10.1016/j.measurement.2017.10.023

52. Rusnák M, Sládek J, Pacina J, Kidová A (2019) Monitoring of avulsion channel evolution and river morphology changes using UAV photogrammetry: Case study of the gravel bed Ondava River in Outer Western Carpathians. Area 51(3):549–560. https://doi.org/10.1111/area.12508
53. Sharma L, Garg PK (eds) (2023) Technological prospects and social applications of society 5.0, 1st ed. Chapman and Hall/CRC. https://doi.org/10.1201/9781003324720
54. Scorpio V, Comiti F, Liébault F, Piegay H, Rinaldi M, Surian N (2024) Channel changes over the last 200 years: A meta data analysis on European rivers. Earth Surf Process Landf. https://doi.org/10.1002/esp.5848
55. Shahtahmassebi AR, Li C, Fan Y, Wu Y, Lin Y, Gan M, Wang K, Malik A, Blackburn GA (2021) Remote sensing of urban green spaces: A review. Urban For & Urban Green 57:126946. https://doi.org/10.1016/J.UFUG.2020.126946
56. Shojanoori R, Shafri HZM (2016) Review on the use of remote sensing for urban forest monitoring. Arboriculture & Urban Forestry (AUF) 42(6):400–417. https://doi.org/10.48044/JAUF.2016.034
57. Svetozarov N, Gallay M (2024) Comparison of 2D and 3D approach for assessing the amount of tree vegetation in the city of Košice using airborne laser scanning data. Geograficky Casopis 76(3):221–247. https://doi.org/10.31577/GEOGRCAS.2024.76.3.12
58. Ullman S (1979) The interpretation of visual motion, vol 229
59. UN/ISDR. (2009). UNISDR terminology on disaster risk reduction. Retrieved from www.unisdr.org
60. Wyżga B, Zawiejska J, Radecki-Pawlik A (2016) Impact of channel incision on the hydraulics of flood flows: Examples from Polish Carpathian rivers. Geomorphology 272:10–20. https://doi.org/10.1016/j.geomorph.2015.05.017
61. Xu B, Xiao X, Wang Y, Kang Y, Chen Y, Wang P, Lin H (2025) Concept and framework of digital twin human geographical environment. J Environ Manag 373:123866. https://doi.org/10.1016/J.JENVMAN.2024.123866
62. Zhang J, Zhu J, Zhou Y, Zhu Q, Wu J, Guo Y, Dang P, Li W, Zhang H (2024) Exploring geospatial digital twins: a novel panorama-based method with enhanced representation of virtual geographic scenes in Virtual Reality (VR). Int J Geogr Inf Sci 38(11):2301–2324. https://doi.org/10.1080/13658816.2024.2386064;PAGE:STRING:ARTICLE/CHAPTER
63. Zhao L, Fan X, Hong T (2025) Urban heat island effect: Remote sensing monitoring and assessment—methods, applications, and future directions. Atmosphere 16(7):791. https://doi.org/10.3390/ATMOS16070791
64. Zhou D, Xiao J, Bonafoni S, Berger C, Deilami K, Zhou Y, Frolking S, Yao R, Qiao Z, Sobrino JA (2018) Satellite remote sensing of surface urban heat Islands: Progress, challenges, and perspectives. Remote Sens 11(1):48. https://doi.org/10.3390/RS1101004
65. Sharma L, Garg PK (eds) (2025) Digital twins and simulation technology concepts and applications, 1st ed. Chapman and Hall/CRC. https://doi.org/10.1201/9781003582489
66. Sharma L, Garg PK (eds) (2024) Deep learning in internet of things for next generation healthcare, 1st ed. Chapman and Hall/CRC. https://doi.org/10.1201/9781003451846

# Chapter 8
# Infrastructure Management and Asset Tracking Using Digital Twins for Geo-Spatial Science

**Rajeev Kumar, Balaji Aryan Singh, Mrityunjay Sharma, Shankar Kumar, and Shubham Kumar**

**Abstract** Digital twins, as digital replicas of physical assets that reflect the real-time status of those assets in the context in which they currently reside, promises to reshape how organizations manage their infrastructure (or physical) assets, integrating rich real-time data, geospatial intelligence, and next generation analytics. This chapter examined the architecture, data exchange standards and operational models of digital twins from the perspective of geospatial science and focus on infrastructure assets including transportation networks, utility grids, and urban facilities. This conversation touched on issues of asset lifecycle management through the use of digital twins, employing Internet of Things (IoT) telemetry, GIS integration, and artificial intelligence to identify when to design, construct, operate, maintain and decommission those assets. It also examined the digital twin enabling technologies, including spatial data infrastructures, systems that utilize cloud-edge architectures, and immersive user interfaces to build scalable, resilient, and responsive twin systems. Case studies were presented across diverse sectors of the economy including logistics, railways, port management, and environmental monitoring to unpack potential future applications of these technologies. The chapter also considered technical challenges such as interoperability, security, operation performance, and sustainability, while describing near term trends associated with digital twins such as autonomous digital

---

R. Kumar (✉)
Institute of Technical Education and Research (ITER), SOA University, Bhubaneshwar, India
e-mail: mrrjvkmr@gmail.com

B. A. Singh
Department of Computer Science, ITER, SOA University, Bhubaneshwar, India

M. Sharma
UCL School of Management, London, England, UK
e-mail: mrityunjay.sharma.24@ucl.ac.uk

S. Kumar
Department of Space, Vikram Sarabhai Space Center, ISRO, Trivandrum, India

S. Kumar
World Energy Council India (WECI), New Delhi, India



L. Sharma et al. (eds.), *Digital Twin Technology in Geospatial Science: Trends, Applications and Challenges*, Data-Intensive Research,
https://doi.org/10.1007/978-3-032-20311-3_8

twins, AI-powered decision making, and the development and commercialization of Digital Twin-as-a-Service platforms. Overall, the findings describe significant opportunities presented by digital twins to support smarter, more dynamic, and resilient management of infrastructure within geospatial contexts.

**Keywords** Digital twin · Geospatial science · GIS integration · IoT sensors · Spatial Data Infrastructure (SDI) · Cloud-edge architecture · Predictive maintenance · 4D visualization · Artificial intelligence · Data interoperability

## 1 Introduction

The complexity of modern-day infrastructure networks—encompassing road and rail networks, utilities, urban settings, and industrial plants—requires a new paradigm of monitoring, analysis, and optimization. Traditional asset management approaches are reactive, siloed, and limited in their ability to emulate dynamic real-world settings. As a counterpoint, the conjoining of Digital Twin (DT) technology with geo-spatial science has emerged as a viable working paradigm for real-time infrastructure management and asset monitoring. A digital twin is a current digital model of a real-world asset or system, maintained in real time through continuous data interchange [20–22]. With the addition of location and time data from Geographic Information Systems, remote sensing, and Global Navigation Satellite Systems, digital twins enable very precise tracking of assets in space and time. Digital systems enable simulation, monitoring, and prediction at scales ranging from individual buildings to national transport systems. Asset tracking as infrastructure management is significantly supported by geo-spatially aware digital twins.

Integrating real-time location, usage status, environmental conditions, and structural integrity data into a comprehensible representation of operations facilitates predictive maintenance, risk assessment, and resource allocation. This chapter discusses cross-disciplinary convergence of geo-spatial science and digital twins for infrastructure management and asset tracking [23, 24]. This chapter presents fundamental concepts, architectural principles, and practical applications, including practical case studies and a Python-based prototype simulation. The chapter aims to instill researchers and practitioners with a deep appreciation of this novel field of inquiry and invite further investigation of spatially intelligent infrastructure systems.

## 2 Foundational Concepts

A Digital Twin (DT) is a virtual representation of a physical thing, process, or system continuously updated using real-time data from sensors and monitoring devices. It builds upon what is often static 3D models by adding a wealth of real-time telemetry, behavior modeling, and predictive analysis. According to the Digital

Twin Consortium [1, 3] a digital twin is also considered "a virtual representation of physical entities and processes, synchronized at specified frequency and fidelity". This is a powerful definition, describing the dynamic nature of digital twins, which distinguishes them from static forms or more typical CAD systems.

## *2.1 Definition and Scope of Digital Twins*

### (a) **Core Components of a Digital Twin System**

A functioning digital twin system relies on a set of interdependent variables, every one of which contributes to the performance of the system in the degree to which the model will simulate and respond to its real-world counterpart. The core of the system is the physical one—the actual asset or infrastructure component such as a bridge, pipeline, or building. The physical component is the ultimate source of information and the real-world behavior to be simulated by the digital system. Its performance under different conditions, structural condition, and interaction between it and the external environment are all the inputs that form the virtual twin.

Alongside the physical, we now have a virtual model which serves to replicate the physical asset's structure, behavior, and life cycle. Unlike static schematics or discrete virtual models, this one stays up to date (i.e. it is constantly computing new values). It receives and maintains pace with the physical by using real-time data arriving from in-place sensors, telemetry streams, etc. The systems data interface layer is the connection that maintains the two systems together by importing data from many sources, including IoT devices, wireless sensors, SCADA systems, and telemetry programs. In summary, the virtual model not only looks like its counterpart, but it behaves in the same way, reacting to changing input in the moment.

Once the data is captured, it needs to be used, and that is where the processing and analytics engine comes into play. This software does everything from basic condition monitoring all the way up to more advanced things like predictive modeling, simulation, and forecasting using machine learning. It allows the system to determine what can go wrong before it actually does, to maximize performance, and to model different operating conditions without impacting the actual asset. To put it all together, the visualization layer and interface give people the tools to interact with the twin. From simple dashboards, GIS visualizations, to even immersive interfaces like virtual and augmented reality, users—planners, engineers, or technicians—can explore the model, model conditions, and make decisions. It is no longer about visualizing the system but understanding it, reacting to it, and optimizing it based on real-world data insights [2, 25–27] (Fig. 1).

### (b) **Problem Solution**

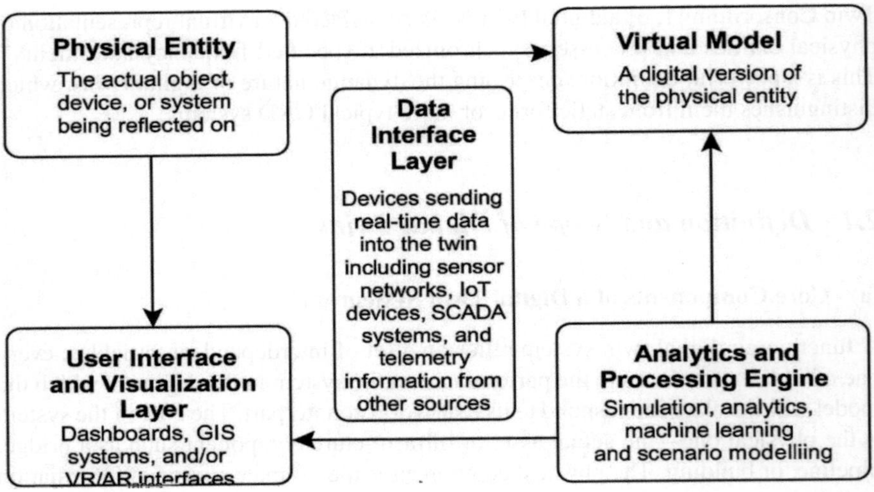


**Fig. 1** Simplified performance model of a critical access module

The notion of a digital twin first appeared prominently in the early 2000s when NASA began to use the concept [3], as a way to simulate and monitor spacecraft systems remotely during mission operations. This case provides an initial use case showing how digital twins have the potential to describe the behavior of complex mission-critical physical systems in real time, facilitate fault detection, risk assessment, and remote diagnostics while based on Earth. Since this early use of the digital twin concept, the digital twin paradigm has advanced a significant amount and has been successfully used in a variety of domains outside of aerospace, including but not limited to product development, manufacturing, automotive systems, industrial automation, and energy production. More recently, digital twins have emerged in the realm of civil infrastructure as the need for real time monitoring, predictive maintenance, and lifecycle management grows.

The speed of change and development of digital twin capabilities has primarily been enabled by various technologies that surround the industry 4.0 revolution. The proliferation of the Internet of Things (IoT) has meant we can instrument infrastructure assets with a dense network of sensors, generating high-frequency data about not only structural performance, but environmental conditions and operational usage. Edge computing is also reducing latency by allowing us to process this data closer to the source, making the data more responsive and reliable in time-critical scenarios. It is now common for Artificial Intelligence (AI) and machine learning [2] algorithms to drive digital twins away from simply being a passive reflection of an object—the trends of today allow for active engagement in behavioral forecasting, anomaly detection and decision optimization. In short, digital twins are no longer experimental technology, but operational systems that are now scalable, intelligent, and agile to respond to all the chaos of modern infrastructure (Table 1).

**Table 1** Comparison with related technologies

| Technology | Purpose | Real-time feedback | Spatial awareness | Simulation capability |
|---|---|---|---|---|
| CAD/BIM models | Design and documentation | X | ☑ | X |
| SCADA systems | Control and monitoring of operations | ☑ | X | X |
| GIS systems | Mapping and spatial analysis | ☑ (limited) | ☑ | X |
| Digital twins | Real-time modelling and forecasting | ☑ | ☑ | ☑ |

### (c) Types of Digital Twin

Digital twins exist in every size and shape, each of which is created to be a match for the level of detail or complexity required to model the physical world. The least complicated is component twins, which would be models of individual components of a system. These might be valves, joints, sensors, or support beams. While small in size, these are often crucial components, and tracking their status in close detail can enable failure detection before it spirals out of control. Tracking components individually is also often the motivation for maintaining the integrity and performance of the overall system in many cases.

A step higher are asset twins, which replicate whole physical assets like turbines, power transformers, or pumps. Asset twins aggregate information from different components to give the whole picture of the asset's condition, use, and efficiency over time. In asset-heavy industries like energy or manufacturing, this whole picture maximizes asset life and minimizes downtime through better maintenance practices. Apart from individual assets, system twins maintain linked components or units in interdependence and interaction with one another. Examples are digital twins of water supply networks, smart power systems, or city public transportation networks. These models help the operators understand how a change in one part of the system can have an impact on the overall system. System-level twins are beneficial to simulate high-scale scenarios—like demand spikes, outages, or climate impacts—which cannot be handled by controlling single parts [5].

Process twins also exist, which do not mirror things or equipment, but how things are accomplished. They mirror workflows, service operations, or business processes—like the patient intake and discharge process at a hospital, or the supply chain at a logistics hub. They help to identify bottlenecks, inefficiencies, or risks in day-to-day operations and serve as a channel to test improvements without disrupting real-world activity.

What is significant is that these kinds of types are not deployed in isolation. Rather, they will be used together. A smart building, for example, may employ component twins for HVAC sensors, asset twins for the heating plant, system twins to bring all building operations together, and process twins to optimize how maintenance scheduling is carried out. Combined, these layers form a robust digital ecosystem that mirrors the physical and functional realities of the environment being cared for.

## 3 Geo-Spatial Science in the Digital Era

### *3.1 Overview of Geo-Spatial Science*

Geo-spatial science is an interdisciplinary science with collecting, analyzing, visualizing, and interpreting spatial data regarding the surface of the Earth and the built environment using instruments and techniques such as satellite-acquired imagery, geographic information systems (GIS), cartography, and spatial statistics.

Increased development in the digital realm has revolutionized geo-spatial science. Geo-spatial science in the digital era utilizes high-resolution satellite imagery with global navigation satellite systems (GNSS), sensor networks, and machine learning to map, monitor, and model spatial phenomena near-real time, adding greater application of knowledge to urban planning, disaster management, infrastructure development, and environmental monitoring.

### *3.2 Role in Infrastructure and Asset Management*

Geo-spatial technologies are the enabling technologies for monitoring and long-term infrastructure asset management. Geo-spatial technologies offer the capabilities required not only to map and monitor assets but also to understand how they behave in space and time. One of the most fundamental applications is infrastructure network mapping. This is the fine-scale representation of road networks, water mains, electricity systems, and telecom cables. With geographic precision, such a map allows planners and engineers to visualize how infrastructure is distributed, where overlaps exist, and how it is linked between points. Besides static mapping, geo-spatial technology also facilitates condition monitoring for a time interval. With the help of high-resolution satellite images or drone images, agencies can perform spatial–temporal change detection to identify structural deterioration, illegal construction, or environmental distress. Monitoring is particularly useful for large structures such as highways, rail lines, or riverbanks, where frequent physical inspections are not practical or are cumbersome.

Geo-spatial science's greatest contribution is possibly its ability to carry out risk simulation. With information regarding land use, topography, climate, and history, geo-spatial models are able to simulate hazard cases [6]. They can, for instance, delineate zones prone to floods, landslides, or earthquakes. If the geo-spatial data are used in infrastructure networks, such information can then be used to assess the risks of structural fatigue and plan preventive measures well in advance of the failure event. Secondly, geo-spatial technologies form the foundation for location-based asset tracking. With GPS or GNSS-enabled equipment, organizations are able to track the actual movement and location of field staff, maintenance fleets, or mobile infrastructure components. This enables better logistics coordination, quicker response to service orders, and better resource utilization in field operations. When linked to a digital twin system, these spatial functions are exponentially more valuable [5]. The digital twin not only graphically represents assets in real time but also spatializes them in their geographic context and appends to them pertinent operating information. This makes it possible to create a twin that is both temporally accurate—displaying current conditions to the minute—and spatially placed, positioning each variation precisely atop its actual location. Asset managers are thus given a richer, more interactive picture of the infrastructure they oversee, allowing them to make more informed, data-driven decisions (ISO 19115:2014).

## *3.3 Spatial Data Infrastructures (SDI)*

A strong digital twin system is dependent on well-organized Spatial Data Infrastructures (SDI). SDI gives the framework for gathering, sharing, and handling spatial data among stakeholders and platforms. Some of the components are (Figs. 2 and 3).

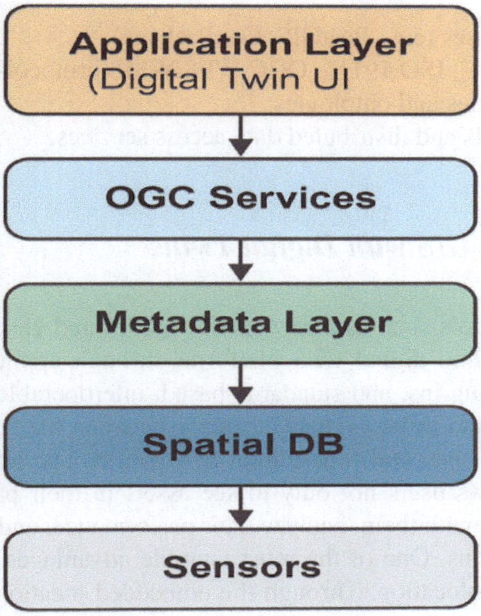


**Fig. 2** Spatial data infrastructures

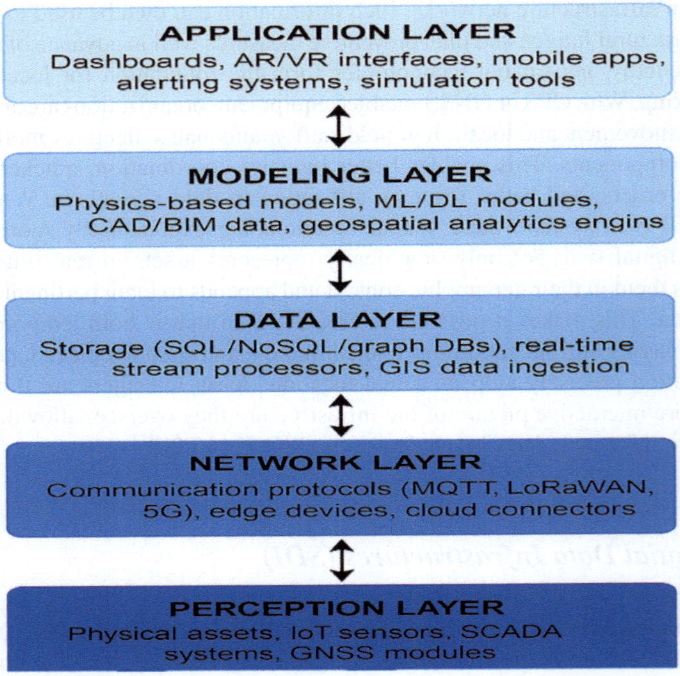


**Fig. 3** Sensor integration and IoT layer

- Geospatial databases (e.g., PostGIS, GeoPackage)
- Data standards (i.e., ISO 19115, OGC WFS/WMS protocols)
- Metadata catalogues and ontologies
- Interoperable APIs and distributed data access services.

## *3.4 Interfacing GIS with Digital Twins*

Sophisticated GIS tools such as Esri ArcGIS, QGIS, and Hexagon GeoMedia are being incorporated into digital twin platforms through application programming interfaces (APIs), plug-ins, and standards-based, interoperable middleware. These connections allow data to be exchanged freely between the spatially enabled GIS world and data-intensive, real-time digital twin world. The integration of the best of both worlds allows users not only to see assets in their physical world but to dynamically interact with them, analyze their performance, and simulate the impact of changing conditions. One of the most tangible advantages of this convergence is real-time asset geolocation. Through the embedded location-sensing sensors on physical assets, asset information can be streamed in real-time into GIS-enabled dashboards to enable real-time spatial monitoring. This is most critical in utilities, transport, and emergency services where situational awareness of the precise location and status of mobile or dispersed assets is the very heart of operational efficiency [8].

Another important skill enabled with 3D GIS integration is 3D city and terrain modeling. Digital twins can be used to create time-series three-dimensional representations of the built environment from geospatial data. Digital twins are not static two-dimensional representations but dynamic simulations that respond to real inputs. They can, for example, be used to monitor urban sprawl, examine shading and air flow patterns around new structures, or simulate traffic flow based on varying infrastructure configurations. Besides, the GIS-digital twin user interface is supported for sensor-based spatial alerts. When an asset crosses a pre-defined spatial boundary (a geofence), or where a sensor detects unusual readings at a location, the system can initiate automated alerts, visual alerts, or invocation of a workflow. Geographic awareness is at the core of anticipatory maintenance, incident response, and risk avoidance.

Further, integration with GIS enables it to execute geospatial queries for simulation management. For example, urban planners can ascertain how a flood event would affect critical infrastructure by querying which roads or utilities would fall within projected flood conditions. Digital twins can then execute simulations for disruption estimation, emergency response planning simulation, or propose rerouting plans based on real-time and historical data [7].

One of the most prominent examples of such integration anywhere is between IBM Maximo and Esri. The solution enables physical asset records in Maximo—asset types, maintenance history, and performance records—to be geospatially attributed and managed with ArcGIS dashboards. With asset intelligence and spatial visualization married together, organizations can have one platform to monitor operations and plan strategy, on real geography.

# 4 Digital Twin Architecture for Infrastructure

## *4.1 System Design Layers*

A typical infrastructure digital twin architecture consists of five integrated layers (Table 2).

## *4.2 Sensor Integration and IoT Layer*

Infrastructure twins rely heavily on **multi-modal sensor networks**. These include:

- **Environmental sensors**: Temperature, humidity, air quality (e.g., for bridges or buildings)

**Table 2** typical infrastructure digital twin architecture

| Layer | Functionality |
|---|---|
| Perception layer | Physical assets, IoT sensors, SCADA systems, GNSS modules |
| Network layer | Communication protocols (MQTT, LoRaWAN, 5G), edge devices, cloud connectors |
| Data layer | Storage (SQL/NoSQL/graph DBs), real-time stream processors, GIS data ingestion |
| Modeling layer | Physics-based models, ML/DL modules, CAD/BIM data, geospatial analytics engines |
| Application layer | Dashboards, AR/VR interfaces, mobile apps, alerting systems, simulation tools |

- **Vibration/strain gauges**: Structural health monitoring
- **GNSS trackers**: Real-time asset geolocation
- **Smart meters**: Utilities (e.g., electricity, gas, water).

Sensor data in this context is often transmitted to edge devices or cloud platforms for preprocessing, mapping and analytics. Also, **Edge AI** components can be used to filter anomalies and reduced data latency (Fig. 3).

### *4.3 Data Models and Standards*

Interoperability is essential for realizing digital twin implementations especially within complex infrastructure systems that draw from many technology and data-derived sources. To support interoperability and effortless communication within a digital twin ecosystem, consistent data models and data formats are necessary. CityGML can be described as the most standardized and practiced data model existing today. CityGML is a 3-dimensional city modeling standard data model representing built environments (e.g., buildings, infrastructure, lands) with multiple layers of representation detail and applications ranging from urban planning to disaster simulation [11].

Another standard is IFC (Industry Foundation Classes) a standard for describing metadata about assets such as buildings and infrastructure. While the IFC standard was originally designed for building information models, the ISO standard is of growing interest from digital twins because it describes physical parameters and materials properties. As commercial software applications evolve to facilitate interoperability, we will likely see many opportunities over the next few years to utilize the metadata definitions afforded by ISO standard IFC. For a more lightweight geospatial data exchange, formats like GeoJSON and TopoJSON are applied. These formats are tailored for web applications and cloud approach and are an efficient way to strip down spatial features like roads or property lines or utility lines without data bloat.

The SensorThings API enabled by the Open Geospatial Consortium specifications provides a schema to manage and query sensor telemetry and data streams within a spatial context. When the SensorThings API is deployed, IoT sensor data can be coordinated and folded into the geospatial analytics of the digital twin. Data models and schemas using standards such as the SensorThings API can create more consistency across heterogeneous parts of the digital twin, enhance data integrity, and promote integration with legacy architectures. Standards and schemas can form a common base language to share information, which is critical for scaling digital twin systems and collaborating between unrelated organizations.

### *4.4 Cloud-Native and Edge Architectures*

A hybrid cloud-edge architecture is increasingly recognized as the most effective architecture for today's digital twin platforms, as it provides healthy balance, scale, and resilience that supports management of complex infrastructure. The cloud layer supports centralized governance, data aggregation, advanced analytics, and coordinated global priorities, provides the environment to run advanced simulations, train machine learning models, and maintain records encompassing timespan for large spectral and distributed infrastructure networks, and creates the basis for analytics, planned action, and possibly transactional platforms through due diligence recording and/or integration with back-end management systems. Edge computing is important for improving responsiveness and reliability on the asset side of the cloud-edge arrangement. Also, edge devices can enhance latency performance, limit bandwidth, and even operate without action from the cloud for periods of time. The relationship between centralized intelligence and local autonomy allows digital twin systems to be effective in real-world situations where scale and immediacy are crucial aspects [12].

A number of leading technology platforms have leveraged this hybrid model of digital twin deployment. Examples of such solutions are Azure Digital Twins, Siemens MindSphere, Autodesk Tandem, and Amazon Location and SiteWise, which support cloud-native solutions with edge computing resources. These platforms provide a variety of digital twin toolkits to build and manage digital twins—ingestion of telemetry, 3D visualization, geospatial analytics, alert systems, and more. Cloud-edge architecture aids organizations with building digital twin ecosystems that are scalable and resilient to reflect new operations or advancing technology.

# 5 Integration of Digital Twins and GIS

## 5.1 Fusion of Spatial Data with Digital Twins

The integration of Geographic Information Systems (GIS) with Digital Twin (DT) platforms allows a seamless spatial portrayal of physical assets and infrastructure environments. GIS-integrated Digital Twins incorporate a set of spatial attributes that previous models treated as secondary, making space the**primary focus of analysis.**

The integration incorporates:

- Locating physical assets to their geospatial coordinates (i.e., longitude, latitude, elevation)
- Storing sensor telemetry data with historical spatial–temporal databases
- Accommodating contextual information (e.g., weather, topography, land use) to establish situational awareness

**Example**

A water pipeline digital twin with GIS layers applied may represent a water pressure loss due to elevated terrain features and identify geographies of possible pipe leakage via spatial anomaly detection.

## 5.2 Real-Time Visualization and 4D Modeling

Digital twins that can connect with Geographic Information Systems can not only provide a three-dimensional representation of an object or a space but also a four-dimensional experience, adding an active time dimension to traditional 3D geometry. The use of spatial information with time allows for a more robust understanding of how infrastructure assets function, change and degrade over time. This modeling allows not only to see the configuration and location of the asset but also visualize and analyze changing conditions through both spatial and temporal dimensions. For instance, with time-series records, engineers and asset managers can visualize the structural deterioration of an asset and build information about than an asset can deteriorate in advance of minor issues morphing into major challenges [12].

Four-dimensional digital models do more than just allow for retrospective tracking. They can also utilize the past to simulate how assets might behave under a variety of conditions in the future. For example, they can model the stress of increased traffic loads, a shift in environmental conditions, or any other operational demands. This kind of analysis allows planners and operators to have key insights for uncomplicated analysis in advance of their decision-making and planning processes. Real-time geolocation of an array of fixed and mobile assets, such as buildings, bridges, maintenance vehicles, or inspection drones, is another systematic capability this technology has in the operational space. This means that asset managers can have

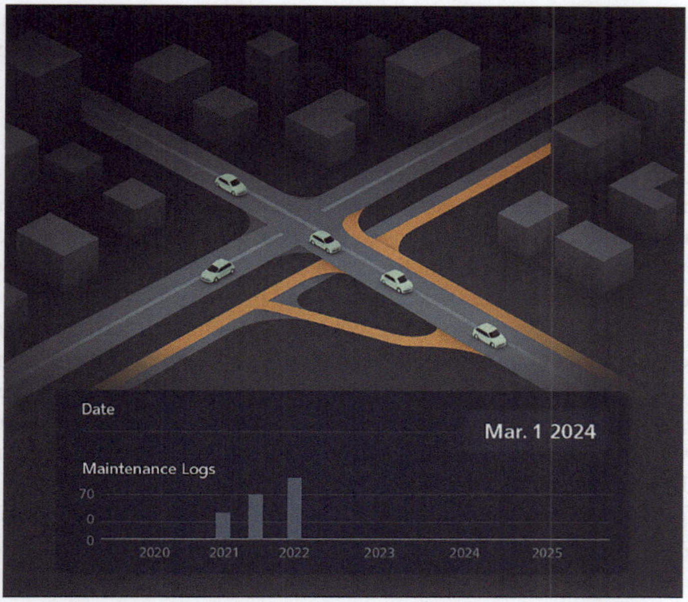


**Fig. 4** Four- dimensional visualization of a digital twin of a road network, showing real-time vehicle location, historical maintenance logs and predictive deterioration

a real-time, live perspective of the situation of operations on the ground and have access to the best knowledge while making any informed decision, such as whether a routine inspection needs further investigation or if an emergency needs to be dealt with.

Several sophisticated data tools allow this type of visualization and analysis to occur. Systems such as CesiumJS, Bentley iTwin, and ArcGIS GeoBIM allow us to construct and navigate through four-dimensional digital representations of the built infrastructure, connecting real-time sensor and data from operations with 3D models and historical data records. These platforms come with interactive tools that include elements such as timeline sliders, in which the user has the ability to navigate between the past, present and expected future states, capturing a truly dynamic view of the built environment (Fig. 4).

## 5.3 *Spatial Queries and Geo-Analytics*

In geo-integrated digital twin systems, spatial queries and geo-analytics are techniques for asking questions and subsequently understanding infrastructure relative to its environmental and operational contexts. These spatial queries and geo-analytics allow users to pose targeted questions that will yield relevant information and aid in making decisions. For example, in an instance an operator can ask about 'Which bridges are within five kilometers of a seismic fault lining?' This may help identify bridge for managing prioritized inspections or retrofits based on enhanced levels of risk. Similarly, an asset manager can ask, which substations fall within the modeled floodplain or buffer zone? This may help with emergency response planning [13] or planned reinforcement. A further use case may include identifying underground cables that may be impacted by a planned road expansion so that planners can mitigate impacts and coordinate any alterations to the infrastructure before they occur.

The benefit of spatial queries includes not only understanding where assets exist but also how they relate to their surroundings and to each other. Geospatial analytics tools take this a step further by applying techniques, such as buffering to model zones of influence, interpolation to estimate values across space where no direct measurement exists, and heatmapping where concentrations of risk or activity are visually expressed, to facilitate risk modeling, resilience assessments, and in designing infrastructure that can both withstand, and adjust to, changing environmental conditions. When embedded within the digital twin, these analytical tools can remove an organization from a static map to a dynamic, data-driven model with operational and strategic value added.

## *5.4 Artificial Intelligence for Spatial Intelligence*

Artificial intelligence, particularly as it applies to machine learning, is playing a more crucial role in enhancing the spatial intelligence of geospatially enabled digital twin technologies. Utilizing ML algorithms with geospatial data empowers digital twins to go beyond simple post hoc monitoring to also provide foresight that is predictive, adaptive, and contextual to its environment. One of the most notable uses of ML is the identification of anomalies using imagery from passive and active data sources, such as thermal sensors, webcams, and drones. For example, a machine-learning model can analyze thermal imagery to detect insulation failures in buildings, pipelines, or electrical systems that would generally not be detected until a much larger issue has arisen [6].

Another beneficial application of AI in this context is using object detection algorithms in drone-acquired imagery. These models could identify and classify physical changes to a large physical network—appearing new school buildings, erosion along a bank, or damage to infrastructure system after a meteorological event—autonomously. This has obviated a manual inspection and enabled faster and data-driven responses.

In addition to current condition detection, machine learning can be used to forecast future conditions the asset may face by analyzing how environmental factors interact with patterns of space use over time. For example, if a digital twin of a rail system could learn how temperature variations, moisture levels, and frequency of trains contribute to the degrading of track, it would be able to better estimate when each of those conditions would lead to needing routine maintenance. Through the employment of AI into GIS-enabled digital twins, the organization not only gains the ability to visualize and monitor infrastructure, but now they can anticipate how to manage the assets in real time, according to changes in conditions and patterns that develop in the world.

**Case Study 2: GeoAI in Pipeline Monitoring**

An oil company in Canada used a GIS-integrated DT with satellite data and AI to detect vegetation anomalies near pipelines, indicating possible underground leaks. Detection accuracy exceeded 92% using multi-temporal data analysis.

# 6 Asset Tracking and Lifecycle Management

## *6.1 Lifecycle Phases of Infrastructure Assets*

Every infrastructure asset has an individual lifecycle. It is from the time it is conceived until it is retired or renewed. Generally, the lifecycle begins with the design and construction of the asset. It is at this point that the physical and functional characteristics of the asset have been defined and have become a physical structure. Digital twins are valuable at this stage by inserting the particulars of the asset—specifications, design models, and materials—into a virtual twin that can be accessed throughout the lifecycle of the asset. When the asset enters the operation and monitoring phase, the digital twin will transition from being a passive record of the static design to a dynamic, real-time reflection of the asset's performance and condition. Sensor data, telemetry systems and external environmental inputs proceed into the twin, so that operators are aware of the usage patterns, can note premature degradation and optimize the operational accuracy on a day-to-day basis.

As the asset ages over its lifecycle to ultimately require some maintenance and repair, the digital twin, in this case, is extremely useful as it will collate historical information, currently available data on asset performance, and predictive models to intervene at the right time and place. The digital twin will be able to hone in on maintenance activities, be more targeted in managing the resource to perform the maintenance or repair and keep asset downtime to a minimum [13]. Eventually, and at some point, an asset reaches the point of decommission or upgrade. Digital Twins will assist this stage as they provide a full record of the history of an asset, including all of the design intents, modifications, in-use periods, and also condition, over an asset's life cycle. With this information, the digital twin aids in the decision to continue with a replacement versus refurbishment blend of actions, that represents the most respectable cost and sustainable option. At all points of the assets lifecycle, to strengthen asset management, you have specifications that are embedded, live data that describe performance, the added spatial context of rich data, that includes what the asset looks like in reality, and all of this makes the digital twin integral to sustaining reliability and longevity in our infrastructure.

**Table 3** Modern asset tracking integrates various technologies

| Technology | Function | Example |
|---|---|---|
| RFID | Short-range passive identification | Inventory and warehouse management |
| GPS/GNSS | Real-time geolocation | Mobile equipment, fleet management |
| BLE Beacons | Indoor positioning and proximity | Hospital infrastructure tracking |
| QR/NFC tags | Physical-digital linkage | Field asset inspection with smartphones |

**Table 4** Digital twin of an urban street lighting system: parameters, traditional maintenance and predictive maintenance

| Parameter | Traditional maintenance | DT-enabled predictive maintenance |
|---|---|---|
| Frequency | Scheduled | Condition-based |
| Downtime | High | Low |
| Asset failure risk | Reactive | Proactive |
| Data dependency | Minimal | High |

## *6.2 Asset Tagging and Identification*

These identifiers are linked to asset records within a DT environment, creating a **digital thread** from the field to control centers (Table 3).

## *6.3 Predictive Maintenance and Condition Monitoring*

Combining assets telemetry, machine learning, and artificial intelligence bring predictive maintenance to a new level that drastically reduces downtime and increases the lifecycle of assets.

**For Example**

A digital twin of a railway track combines vibration sensor and thermal imaging sensors to predict fatigue on rail. A digital twin of an urban street lighting system combined with energy consumption data can predict when light bulbs will fail (Table 4).

## *6.4 Tracking Mobile Infrastructure Assets*

Tracking mobile infrastructure assets is a vital capability included in many industries, including logistics, construction, and utilities, where knowing the exact location and state المعارضةmoving equipment or vehicles directly impacts operational efficiency and safety in supply chain processes. Whether delivery trucks, mobile substations, service or utility vans, or construction machinery, using digital twins to track moving assets in real time can aid in managing operations across combinations of industries. One of the greatest advantages is better routing efficiency. Constant updates before the mobile asset paradox occur and location updates can facilitate more robust routing between spatial and operational contexts to save fuel, prioritize time-sensitive deliveries, and reduce congestion.

In addition to day-to-day activity improvements, digital twins will have usefulness in response avoiding future disasters that are needless and planning for responses workplaces. In the case of emergency situations or urgent maintenance dealing with infrastructure, when a mobile source of energy must be used inadvertence, custodian, it is easier to call what asset can mobilize or not with real-time, precise data to move people more efficiently to respond to service issues identified by faults in the infrastructure. This can be especially pertinent in utility networks where infrastructure may be impacted by storm events, or service companies when heavy machinery must move quickly and coordinated with little notice.

Another key advantage is theft and loss prevention. Digital twins powered by telematics allow near real-time tracking of valuable equipment, including the use of GPS coordinates [14], measurements from accelerometers, and onboard diagnostics. Telecommunication capabilities allow for geo-fencing, which means setting perimeters around a workspace. If one component of your mobile infrastructure and fleet leaves the defined geo-fenced area, alerts can automatically notify you of a potential unauthorized ride or theft. In this sense, the digital twins facilitate organizational knowledge about their mobile infrastructure's location and status, allowing for better protection, efficiency, and resilience.

**Case Study 3: Smart Utility Fleet Tracking**

In India, a power three utility can keep track of their service vehicles from digital twins. The GPS tracking was integrated into their GIS-DT and used the location services to improve route efficiency on the fly as crews worked to restore power, improving the restoration time by 35% for outages during monsoon periods.

## 7 Use Cases and Applications

### *7.1 Smart Cities and Urban Infrastructure*

Digital twins integrated with GIS are foundational for smart city programs and allow for centralized control of infrastructure that is critical to the safe and efficient operation of cities.

Key applications include:

- Monitoring of roads, traffic lights, waste collection, and utility use in real time.
- Emergency response modeling (particularly for things like modeling fire spread or traffic diversion from accidents).
- Dynamic congestion pricing and urban mobility optimization.

**Case Study 4: Singapore Virtual City**

The Singapore Land Authority developed a 3D city model for the nation that incorporates digital twins of urban infrastructure. The system provides a platform for urban planning and energy modeling. In particular, it allows for simulation of the impact of future development projects on air flow and sunlight access.

## *7.2 Transport and Logistics*

The use of digital twins continues to impact the port, rail, and logistics industries by improving real-time tracking and operational performance. In port management systems, digital twins are monitoring containers and vehicles in real time and are designed to improve docking processing times by optimizing the load/unload schedules. Reducing delays allows the port agency to gain operational efficiencies and improve throughput throughout the terminal. The delivery of real-time information makes it possible to track and monitor the state of rail transport assets, provide condition-based maintenance of the rail stock, and identify possible fatigue failure areas on the rail track networks by means of spatial heat mappings [15]. Logistics agencies are employing digital twins to simulate optimal routes for delivery vehicles according to live traffic and environmental data, while allowing low-emission and fuel-efficient operations. The examples provided outline the utility of digital twins in assisting industries to achieve better and more sustainable options going forward.

## *7.3 Energy and Utilities*

Digital twins are becoming increasingly prevalent in the energy and utilities sector, where they offer operators real-time situational awareness about how an asset is performing. With electric grids, for instance, a digital twin can help identify situations, such as phase imbalance, transformer overheating, or line sag, and inform the operator of the situation permitting them to resolve it before the asset experiences a larger failure. For water treatment facilities managers can use the digital twin to model scenarios, complete with aspects of chemical dosing, fluid flow, and the overall structure of the facility and integrity, with operating information from sensors to operationally respond. In renewable energy applications, such as with wind farms or solar farms, operational and geospatial data (weather, shading potential by terrain, etc.)

was leveraged with the digital twin to validate or estimate output. These examples show the potential for digital twin technology to improve the reliability, efficiency, and sustainability of vital infrastructures.

## *7.4 Construction and Civil Engineering*

Digital twins are considered valuable tools in the construction and civil engineering industries, providing innovative methods to improve project oversight, planning, and the long-term performance of a civil asset. Construction companies are increasingly using digital twins to understand the location and operational status of semi-permanent heavy equipment at building sites. This provides management with important enough information to ensure that their machinery is properly deployed, maintained, and secured. These integrate systems improve safety and productivity on-site, manage materials, and monitor equipment flows.

A significant feature of digital twins is the ability to layer Building Information Models (BIM) onto GIS-based terrain data. That allows construction teams to think about staging and construction sequencing in context with their geographical surroundings. This is particularly helpful in allowing project teams to identify concurrent completion operations, help avoid project conflicts, and allow for better decision making in response to changes that occur due to site conditions as the construction schedule evolves. Digital twins also convey value after construction is complete by facilitating the evaluation of all post-construction effects, such as settlement, stress, and structural performance. Data collected by geotechnical sensors that measure strain, pressure, tilt, and displacement, provided to the digital twin functioning in real time provide the owner/operator the ability to assess how the construction asset is responding to their environment and operations over time. Having both the design intent monitored and the as-built situation managed during the entire structure lifecycle is a significant contribution by digital twin systems [16] (Fig. 5).

## *7.5 Environmental Monitoring*

Digital twins are developing so that they can be useful in environmental monitoring, synthesizing real time geospatial data, sensor feeds and prediction models, and serving as a mechanism to evaluate potential risks. One of the best-known examples of this is with wildfire risk management. In this framework, variables like, vegetation maps, real time weather telemetry, and topographic modeling are synthesized to forecast how fires may behave and spread in a given region, allowing planners to better prepare mitigation and emergency response plans. Digital twins have also

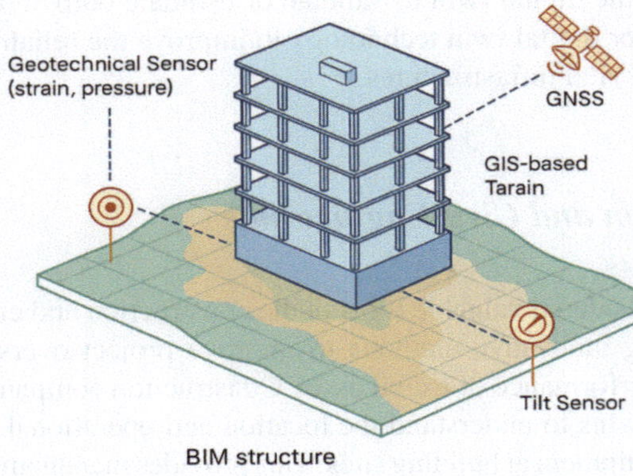


**Fig. 5** A BIM structure superimposed on GIS terrain with labeled sensors (strain, pressure, GNSS, tilt sensor) and network connections

been used in landslide early warning systems, combining slope movement, soil moisture maps, and detections of the risk of landslide failures, sometimes even alerting decision makers of the risk prior to the landslide failure.

Digital twins have been used for air and water pollution monitoring, where the spatial and temporal mapping of emissions, for example of pollution resulting from vehicular traffic or industrial sites, assists managers in understanding the potential impacts of the pollutant, and where their interventions could be most effective [17]. The foregoing use cases highlight the significant role that digital twin applications have in improving smarter, faster responses to environmental conditions.

# 8 Challenges and Limitations

## *8.1 Data Interoperability and Standards*

One of the biggest problems of digital twins for infrastructures is pulling together heterogeneous data inputs from different systems. It seems very little attention has been paid to how interoperate inputs from the legacy CAD (Computer Aided Design) drawings, real-time sensor streams, satellite imagery with differing map projections, SCADA (Supervisory Control and Data Aquisition) proprietary formats etc. into one digital twin. These data types have many differences which mean that when you attempt to combine, align or analyze them in some way in a single digital twin you often find differences at the semantic level. In the absence of shared formal representations (for example, standardized formats, schemas and ontologies), we ran into various semantic inconsistencies and issues with alignment which often compromised our ability to make sense of the data in a consolidated manner and

**Table 5** Mapping of standards to data types

| Standard/format | Associated data type/purpose |
|---|---|
| IFC (Industry Foundation Classes) | BIM and building/infrastructure metadata |
| CityGML | 3D urban models, city-scale digital representations |
| GeoTIFF | Raster data, satellite imagery, elevation models |
| SensorThings API (OGC) | Real-time IoT sensor streams, telemetry data |
| GeoJSON/TopoJSON | Lightweight vector data for spatial features |
| LAS/LAZ | LiDAR point cloud data |
| GML (Geography Markup Language) | Complex geospatial data exchange |
| WMS/WFS (OGC protocols) | Web-based map and feature services |
| Shapefile | Legacy vector geospatial data |

render it meaningfully. Given the amount of differences between and across very different data types there must surely have been thousands of potential semantic mismatches [18].

Digital twin platforms are typically anchored to a well-defined set of data standards, which map to well-defined data types and also data usages, in order to help us navigate some common interoperation and integration of otherwise incompatible systems (Table 5).

## 8.2 *Scalability and Performance*

When using digital twins to assist with large infrastructure systems composed of many thousands of interconnected assets, it is important to make considerations around scalability and performance. The complexity and amount of data produced in these situations can be significant, since digital twins operate on frameworks of scalability to ensure performance through their use of spatio-temporal indexing in databases. Spatio-temporal data sources employ scalable databases that can manage a large amount of spatial data connected to streams of time-series. Databases that support spatial data such as PostGIS and MongoDB with geospatial extensions are capable of efficiently storing, indexing, querying, and retrieving the spatial data. Alongside spatial data storage databases, digital twins often make use of stream processing frameworks such as Apache Kafka and Apache Flink that allow the twin to operate on high-frequency data in real-time, processing events, predictive analytics, or automating events. Visualization should also be a point of performance concern, especially for digital twins that render 3D or 4D models and provide dynamic, live updates.

Engaging visualizations can also be made interactive through the use of performance platforms such as Unreal Engine and Cesium, to provide high-fidelity experiences without exhausting resources and performance capabilities. Beyond the limits of technology, digital twins can still reach bottlenecks of resources, whether operated in a cloud computing environment or at the edge of infrastructure assets, that can limit system performance. Bottlenecks caused by resource limits can slow the speed of data ingestion, data analysis, or data visualization to the point that the digital twin has poor responsiveness and is nearly worthless for purposes of time-sensitive decision making. Therefore, technology to ensure scalable systems that are capable of performance must be built into the design and tooling of a digital twin for long-term success [18].

## 8.3 Real-Time Data Accuracy

Real-time data accuracy is a basic hurdle in the operation of digital twins. The reliability of the twin's outputs is wholly dependent on the quality of data it receives. The operational stream of data can be compromised by several issues. For example, sensor drift and sensor malfunction can lead measurements to drift increasingly from true measurements, causing potential errors to go unnoticed until the deviations become stark between expected predictions, or visualizations. Second, telemetry can also demonstrate a degree of accuracy impairment through environmental interference, system degradation, and communication failures that can lead to delays and reliability issues with the input data streams.

In urban environments or with complex infrastructure, loss or interference in GPS tracking signals (including secure connection interruption when moving through tunnels and dense building coverage) can impair the spatial accuracy of location tracking and subsequent spatial analytics. If a digital twin is to deliver utility to an operator, then it must be designed with mechanisms to guarantee the trustworthiness of the data and telemetry streams that inform the twin. This could include installation of redundant sensor streams, automated validation and data filtering, and algorithms that detect and adjust for deviations. In this regard, the digital twin can provide a useful and dependable reflection of the real-world context.

## 8.4 Cybersecurity and Privacy

Considering that the digital twin system may call exposures such as APIs and control interfaces as well as sensitive geolocation, cybersecurity and privacy represent significant risks. Digital twin systems are vulnerable to breaches that provide attackers the ability to utilize the real time telemetry, expanded asset metadata, and control capabilities that extend the attack surface; improperly managed, attack will exploit remote monitoring and other weaknesses, posing serious risks. One area of concern is obtaining unauthorized access to data, operational data, or control systems—presenting risks to continuity, as well as proprietary information theft.

The other is the integrity of the underlying geolocation or sensors, where the integrity of the twin relies on accurate readings, validated modes, or commands issued to physical assets are actually executed for the malicious actor's purpose. Privacy risks include exposing patterns in maintenance, service or operational activity that exposes operational vulnerabilities as well as opportunities to target operations. Best practices in addressing cybersecurity/ privacy risks in digital twin platforms include implementing strong encryption of data at rest and in transit, implementing fine-grained access control to limit user privileges, and implementing "zero-trust" architectures whereby no user and devices are not trusted by default until validated. These practices, as a combination of policies, assist in the prevention of digitally compromising the integrity, confidentiality, and availability of digital twin platforms.

## 8.5 Cost and Implementation Complexity

The cost and complexity of implementing digital twin technology represents a major drawback to the widespread adoption of digital twin technology, particularly for major public infrastructure projects that operate under strict budgets and scarce resources. The initial investment for sensors, modeling platforms (which could also include purchasing or developing new modeling platforms), training, and integrating all of this with current GIS and asset management tools will depend not only on the reporting requirements but will not be small.

Additionally, implementation often requires complex coordination across many stakeholder groups from technology providers, governments, planners and operational groups—if this coordination is lacking then projects risk being delayed, increased scope of work or loss of function. Gaps in technical skills may slow down implementation for digital twins, regardless whether they are with data science, geospatial data analysis, or systems integration, these can increase the time and reliance to put into technology, or can require investment in building capacity for the user group. Addressing these barriers will demonstrate the promise of digital twins as long-term planning and management tools for smarter and more resilient infrastructure.

**Case Study 5: Failed Deployment of Municipal Twin**

A European city's plan to twin its metro system was halted after encountering data format inconsistencies, vendor lock-in, and unresolved legal issues related to public surveillance imagery.

# 9 Future Trends

## 9.1 Integration with AI and Autonomous Agents

One of the most exciting areas for digital twin technology is in conjunction with AI and autonomous agents. As AI systems become increasingly sophisticated, digital twins are transitioning to be more than passive data mirrors and instead, are becoming active decision-support and control systems that can reason under uncertainty. The scope for disruption is even broader in geospatial infrastructure where complex environments and evolving conditions must be considered to arrive at the appropriate adaptive solution. For example, a digital twin that is automated by an autonomous drone or robot can continuously inspect the assets on behalf of humans at a location and can identify damage, gather data about the state (drones capture imaging data), and even take action without a human operator on site.

Adaptive infrastructure systems can similarly respond to live conditions in a complex infrastructure change using methods like reinforcement learning (e.g., traffic light systems that autonomously adjust to traffic conditions). In utility networks, predictive fault detection with AI-enabled digital twins will create opportunities for auto-remediation to allow for smaller faults in a grid to be recognized and fix themselves without human intervention if possible. Together, these developments signal a future in which digital twins are not just passive representations of the physical world but are involved in helping to functionally implement its autonomous operation [19] (Fig. 6).

**Fig. 6** Emerging trend

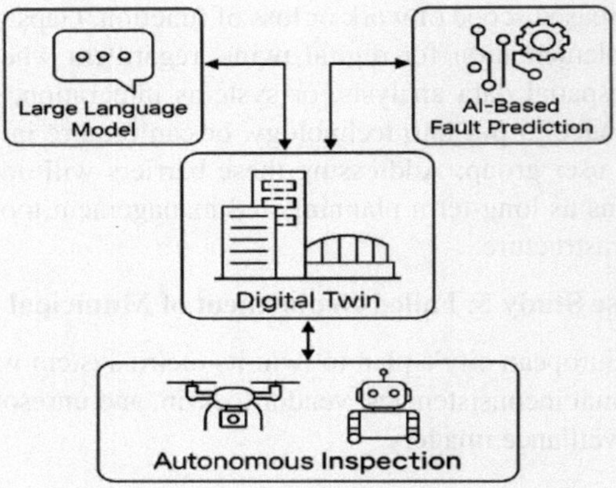

## 9.2 Immersive Interfaces (AR/VR/XR)

Immersive interfaces such as augmented reality (AR), virtual reality (VR), and mixed reality (MR) are fundamentally altering the use of digital twins for both effective management of complex infrastructure. Immersive interfaces enhance spatial awareness and situational awareness by allowing users to engage with digital representations of physical environments more naturally. For maintenance planning with existing infrastructure, for example, VR headsets allow engineers and planners to visualize existing buried utilities or subsurface assets, such as pipelines or buried cables, without conducting a site survey by going through the ground. Similarly, AR offered through smartphones or tablets enhances the ability of field workers to locate and identify physical assets, or assets in real time by combining the locations obtained through GNSS positioning within the Building Information Modeling (BIM) model of the asset. It allows for the location and relevant metadata of assets to be placed directly in the user's view of the physical environment. MR enhances this capability by including additional features for remote inspection and collaboration. Using MR headsets, users PDT can interact with other users in spatially anchored avatars and share digital annotations tied compressors for the Children consideration.

Ultimately correlational policymakers drifting feet from the act suit can efficiently relay the actions to be integers. In many ways, these immersive interfaces encompass illustrated conditions and greater collaboration for the ongoing management of infrastructure contribute to increased accuracy and efficiency with using email based. At eagerness are opening new possibilities for training, safety, and collaborative technological context, supportive collaboration work channels blow.

## 9.3 Blockchain for Asset Provenance and Data Integrity

Blockchain technology is becoming a focal point for identifying mechanisms to improve asset provenance and Richard's data integrity within digital twin systems. Blockchain is a distributed ledger that is tamper-proof and immutable. It could act as a reliable record for the asset's maintenance work, the condition reports of the asset, and the motion history (when, where and for how long)—particularly in sectors such as oil and gas, defense infrastructure, and aviation that require traceability and accountability. An important use case is establishing chain of custody for the critical components—being able to trace parts used in sensitive operations or equipment of value, all the way from manufacture to installation to maintenance. Blockchain has also been considered for recording sensor data integrity which provides a verifiable/auditable record for compliance purposes.

An example being during a regulatory inspection or audit where operators can prove that the streams of data from the digital twin—be it structural sensors or environmental data and monitoring—useful for audit purposes, could not have been altered or manipulated. In sum, having blockchain integrated into a digital twin provides a greater level of trust and accountability, transparency, and reduced probability for data tampering or loss.

## 9.4 Digital Twin-as-a-Service (DTaaS)

Digital Twin-as-a-Service (DTaaS) signifies a meaningful change in the deployment and acquisition of digital twin technology. It provides greater access to a broader audience of organizations as they can now implement digital twin technology in a cloud-native environment. Rather than a consequence of large upfront pricing associated with infrastructure and software costs, DTaaS allows the user to adopt cost modeling and simulation after the fact, only charging for what is consumed. It is now easier for small and medium-sized enterprises and public agencies to adopt digital twin methodologies for asset management and infrastructure. DTaaS platforms will usually provide for cloud support of geospatial data using services like AWS Location and Google Earth Engine, as well as satellite images and spatial analytics, layering these services in the twin environment.

It focuses on providing DTaaS services that are auto-scaling to enable deployments that have many thousands of sensors and devices, which would overwhelm organization resources in practice. Leading organizations in this space are Azure Digital Twins, Siemens MindSphere, ANSYS Twin Builder, and Dassault Systèmes 3DEXPERIENCE. Collectively, these are helping to democratize digital twin technology and offering more flexible, scalable, and accessible options than ever before user [19].

## 9.5 Climate Resilience and Sustainability

In light of increased climate threats, geospatially aware digital twins are becoming increasingly valuable to climate resiliency and sustainability. Governments and public sectors have been using digital twins, along with other geospatial technologies, to model, predict and manage risks related to climate. Digital twins offer the possibility to do things like predicting where coastal zones may flood as sea levels rise so that communities can plan for their defenses, reconfigure land use, and prioritize remediation efforts for their most disadvantaged communities. With cities, systems could be modeled to help planners understand how a heatwave will impact local energy grids; during peak heat events they could view peak loading during a heatwave, while improving overall energy grid reliability [18].

Digital twins help authorities and agencies more closely assess whether transportation infrastructure is sustainable; this may help assess and monitor everything from the carbon footprint of the road connections to the transportation options utilized (for public transport, freight) by its citizens. Geospatially aware digital twins will help advice decision makers on a) historic or current emissions footprints as associated with infrastructures, and even b) aid anticipation and action on the design of future, less harmful, or more resilient infrastructures.

## 10 Code Simulation: Prototype Digital Twin for Urban Asset Tracking

To anchor the theoretical elements, this section uses a simple simulation in Python to model and visualize an asset (i.e., utility vehicle or equipment) movement through a geo-spatial environment in which the asset is continuously tracked through time and conditions are measured in real-time.

Prerequisites:

***Python 3.10+, Folium, Geopandas, Pandas, Random, Matplotlib***

### *10.1 Simulated Scenario*

Track a set of assets (e.g., maintenance trucks) moving through a city grid. Each asset has:

- A unique ID
- Geo-location (latitude, longitude)
- Status (e.g., normal, requires maintenance).

The simulation updates position and status every few seconds, visualized using **Folium** maps (Figs. 7, 8 and 9).

## 11 Conclusion

This chapter has explored the convergence of digital twin technology with geospatial science as a new paradigm for infrastructure management and asset tracking. By building continuously updated virtual representations of physical assets using real-time data, IoT telemetry, and GIS integration, digital twins enable organizations to move from reactive, siloed approaches to predictive, data-driven decision-making. Reflecting the Digital Twin Consortium's definition of digital twins as synchronized, high-fidelity virtual representations of physical systems, this approach offers precise spatial and temporal awareness essential for modern infrastructure networks. Core components such as spatial data infrastructures (SDI), cloud-edge architectures, and

```
import folium  # Library for creating interactive maps
import pandas as pd  # Library for data manipulation and analysis
import random  # Library for generating random numbers
import time  # Library for time-related functions
from folium.plugins import TimestampedGeoJson  # Plugin for time-stamped GeoJSON fea-
tures

# Asset types and simulation configs
ASSET_TYPES = ['vehicle', 'personnel', 'sensor', 'drone']
STATUSES = ['normal', 'maintenance', 'idle', 'malfunctioning', 'charging', 'in transit']
COLORS = {
    'vehicle': 'blue',
    'personnel': 'green',
    'sensor': 'orange',
    'malfunctioning': 'red',
    'drone': 'purple',  # Color for the new 'drone' asset type
    'charging': 'yellow',  # Color for 'charging' status
    'in transit': 'cyan'  # Color for 'in transit' status
}

# Function to initialize assets with random locations and types
def generate_assets(num_assets):
    assets = []
    for i in range(num_assets):
        asset_type = random.choice(ASSET_TYPES)
        lat = 28.61 + random.uniform(-0.01, 0.01)
        lon = 77.21 + random.uniform(-0.01, 0.01)
        asset = {
            "id": f"A{i}",
            "type": asset_type,
            "status": "normal",
            "location": [lat, lon],
            "history": []
        }
        assets.append(asset)
    return assets

# Function to simulate movement and status updates for each asset
def simulate_step(assets, step_num=None, total_steps=None):
    """
    Simulate a single step of movement and status update for each asset.
    If step_num and total_steps are provided, assets will move in a more continuous random
walk.
    """
    for asset in assets:
        # Simulate a random walk with some persistence in direction
        if 'direction' not in asset:
            angle = random.uniform(0, 2 * 3.14159)
            asset['direction'] = angle
        else:
```

**Fig. 7** Codes are shown for reference

```
            # Small random change in direction
            asset['direction'] += random.uniform(-0.3, 0.3)

        speed = 0.0007 if asset['type'] == 'vehicle' else 0.0003
        if asset['type'] == 'drone':
            speed = 0.001
        elif asset['type'] == 'personnel':
            speed = 0.0004

        delta_lat = speed * random.uniform(0.7, 1.3) * random.choice([1, -1]) * random.uni-
form(0.8, 1.2) * abs(random.gauss(0, 1))
        delta_lon = speed * random.uniform(0.7, 1.3) * random.choice([1, -1]) * random.uni-
form(0.8, 1.2) * abs(random.gauss(0, 1))

        # Move in the direction with some randomness
        asset['location'][0] += delta_lat * random.uniform(0.5, 1.5) * (random.random() - 0.5)
        asset['location'][1] += delta_lon * random.uniform(0.5, 1.5) * (random.random() - 0.5)

        # Clamp to a reasonable area (Delhi region)
        asset['location'][0] = min(max(asset['location'][0], 28.60), 28.62)
        asset['location'][1] = min(max(asset['location'][1], 77.20), 77.22)

        # Status changes more likely at the start/end or randomly
        if random.random() < 0.08 or (step_num is not None and (step_num == 0 or step_num ==
total_steps-1)):
            asset['status'] = random.choice(STATUSES)

        # Append timestamped log
        asset['history'].append({
            "coordinates": list(asset['location']),
            "status": asset['status'],
            "timestamp": pd.Timestamp.now().isoformat()
        })
    return assets

# Function to create an interactive map with asset history
def generate_map(assets, step_interval_seconds=5):
    """
    Generate a folium map with animated asset movement using TimestampedGeoJson.
    step_interval_seconds: time between steps in the animation (in seconds).
    """
    m = folium.Map(location=[28.6139, 77.2090], zoom_start=14, control_scale=True)

    features = []
    for asset in assets:
        for idx, record in enumerate(asset['history']):
            # Use a synthetic timestamp for animation, spaced by step_interval_seconds
            synthetic_time = pd.Timestamp.now() + pd.Timedelta(seconds=idx * step_interval_sec-
onds)
```

**Fig. 7** (continued)

```
        features.append({
            "type": "Feature",
            "geometry": {
                "type": "Point",
                "coordinates": record["coordinates"][::-1]
            },
            "properties": {
                "time": synthetic_time.isoformat(),
                "popup": f"{asset['id']} ({asset['type']}, {record['status']})",
                "icon": "circle",
                "iconstyle": {
                    "fillColor": COLORS.get(record['status'], COLORS.get(asset['type'], 'gray')),
                    "fillOpacity": 0.8,
                    "stroke": "true",
                    "radius": 7
                }
            }
        })

    TimestampedGeoJson({
        "type": "FeatureCollection",
        "features": features,
        }, period=f'PT{step_interval_seconds}S', add_last_point=True, auto_play=False,
loop=False, max_speed=5, loop_button=True, date_options='YYYY-MM-DD HH:mm:ss',
time_slider_drag_update=True).add_to(m)

    # Add a legend to the map
    legend_html = '''
    <div style="
        position: fixed;
        bottom: 30px;
        right: 30px;
        width: 210px;
        background: rgba(255,255,255,0.95);
        border: 2px solid #444;
        border-radius: 8px;
        z-index:9999;
        font-size: 14px;
        box-shadow: 2px 2px 8px rgba(0,0,0,0.2);
        padding: 12px 18px 12px 18px;
    ">
        <b>Legend</b><br>
        <div style="margin-top: 6px;">
          <span style="display:inline-block;width:16px;height:16px;background:blue;border-ra-
dius:50%;margin-right:8px;border:1px solid #333;vertical-align:middle;"></span>
          Vehicle<br>
           <span style="display:inline-block;width:16px;height:16px;background:green;border-
radius:50%;margin-right:8px;border:1px solid #333;vertical-align:middle;"></span>
          Personnel<br>
```

**Fig. 7** (continued)

```
        <span style="display:inline-block;width:16px;height:16px;background:orange;border-radius:50%;margin-right:8px;border:1px solid #333;vertical-align:middle;"></span>
        Sensor<br>
        <span style="display:inline-block;width:16px;height:16px;background:purple;border-radius:50%;margin-right:8px;border:1px solid #333;vertical-align:middle;"></span>
        Drone<br>
        <span style="display:inline-block;width:16px;height:16px;background:red;border-radius:50%;margin-right:8px;border:1px solid #333;vertical-align:middle;"></span>
        Malfunctioning<br>
        <span style="display:inline-block;width:16px;height:16px;background:yellow;border-radius:50%;margin-right:8px;border:1px solid #333;vertical-align:middle;"></span>
        Charging<br>
        <span style="display:inline-block;width:16px;height:16px;background:cyan;border-radius:50%;margin-right:8px;border:1px solid #333;vertical-align:middle;"></span>
        In Transit
      </div>
    </div>
    '''
    m.get_root().html.add_child(folium.Element(legend_html))

    return m

import argparse
import sys

def main():
    parser = argparse.ArgumentParser(
        description="Simulate asset tracking and generate an interactive map and CSV export."
    )
    parser.add_argument(
        "-n", "--num-assets", type=int, default=10,
        help="Number of assets to simulate (default: 10)"
    )
    parser.add_argument(
        "-s", "--steps", type=int, default=10,
        help="Number of simulation steps (default: 10)"
    )
    parser.add_argument(
        "--map-file", type=str, default="digital_twin_asset_tracking_map.html",
        help="Output HTML file for the interactive map"
    )
    parser.add_argument(
        "--csv-file", type=str, default="asset_data.csv",
        help="Output CSV file for asset data"
    )
    parser.add_argument(
        "--no-map", action="store_true",
        help="Do not generate the interactive map"
    )
    parser.add_argument(
```

**Fig. 7** (continued)

```
        "--no-csv", action="store_true",
        help="Do not export the asset data CSV"
    )
    args = parser.parse_args()

    print(f"Simulating asset tracking with {args.num_assets} assets for {args.steps} steps...")
    assets = generate_assets(args.num_assets)
    for step in range(args.steps):
        simulate_step(assets, step_num=step, total_steps=args.steps)

    if not args.no_map:
        final_map = generate_map(assets, step_interval_seconds=5)
        final_map.save(args.map_file)
        print(f"Map saved as {args.map_file}")

    if not args.no_csv:
        asset_data = []
        for asset in assets:
            for record in asset['history']:
                asset_data.append({
                    "id": asset['id'],
                    "type": asset['type'],
                    "status": record['status'],
                    "latitude": record['coordinates'][0],
                    "longitude": record['coordinates'][1],
                    "timestamp": record['timestamp']
                })
        df = pd.DataFrame(asset_data)
        df.to_csv(args.csv_file, index=False)
        print(f"Asset data exported to {args.csv_file}")

if __name__ == "__main__":
    main()
import folium  # Library for creating interactive maps
import pandas as pd  # Library for data manipulation and analysis
import random  # Library for generating random numbers
import time  # Library for time-related functions
from folium.plugins import TimestampedGeoJson  # Plugin for time-stamped GeoJSON fea-
tures

# Asset types and simulation configs
ASSET_TYPES = ['vehicle', 'personnel', 'sensor', 'drone']  # Added 'drone' as a new asset type
STATUSES = ['normal', 'maintenance', 'idle', 'malfunctioning', 'charging', 'in transit']  # Added
'charging' and 'in transit'
COLORS = {  # Colors associated with each asset type and status
    'vehicle': 'blue',
    'personnel': 'green',
    'sensor': 'orange',
    'malfunctioning': 'red',
```

**Fig. 7** (continued)

```
    'drone': 'purple',  # Color for the new 'drone' asset type
    'charging': 'yellow',  # Color for 'charging' status
    'in transit': 'cyan'  # Color for 'in transit' status
}

# Function to initialize assets with random locations and types
def generate_assets(num_assets):
    assets = []
    for i in range(num_assets):
        asset_type = random.choice(ASSET_TYPES)
        lat = 28.61 + random.uniform(-0.01, 0.01)
        lon = 77.21 + random.uniform(-0.01, 0.01)
        asset = {
            "id": f"A{i}",
            "type": asset_type,
            "status": "normal",
            "location": [lat, lon],
            "history": []
        }
        assets.append(asset)
    return assets

# Function to simulate movement and status updates for each asset
def simulate_step(assets):
    for asset in assets:
        delta_lat = random.uniform(-0.0005, 0.0005)
        delta_lon = random.uniform(-0.0005, 0.0005)
        asset['location'][0] += delta_lat
        asset['location'][1] += delta_lon

        if random.random() < 0.1:
            asset['status'] = random.choice(STATUSES)

        # Append timestamped log
        asset['history'].append({
            "coordinates": list(asset['location']),
            "status": asset['status'],
            "timestamp": pd.Timestamp.now().isoformat()
        })
    return assets

# Function to create an interactive map with asset history
def generate_map(assets):
    m = folium.Map(location=[28.6139, 77.2090], zoom_start=14, control_scale=True)

    features = []
    for asset in assets:
        for record in asset['history']:
            features.append({
                "type": "Feature",
```

**Fig. 7** (continued)

```
            "geometry": {
                "type": "Point",
                "coordinates": record["coordinates"][::-1]
            },
            "properties": {
                "time": record["timestamp"],
                "popup": f"{asset['id']} ({asset['type']}, {record['status']})",
                "icon": "circle",
                "iconstyle": {
                    "fillColor": COLORS.get(asset['status'], COLORS.get(asset['type'], 'gray')),
                    "fillOpacity": 0.8,
                    "stroke": "true",
                    "radius": 7
                }
            }
        })

    TimestampedGeoJson({
        "type": "FeatureCollection",
        "features": features,
            },    period='PT1M',    add_last_point=True,    auto_play=False,    loop=False,
max_speed=1).add_to(m)

    # Add a legend to the map
    legend_html = '''
    <div style="
        position: fixed;
        bottom: 30px;
        right: 30px;
        width: 210px;
        background: rgba(255,255,255,0.95);
        border: 2px solid #444;
        border-radius: 8px;
        z-index:9999;
        font-size: 14px;
        box-shadow: 2px 2px 8px rgba(0,0,0,0.2);
        padding: 12px 18px 12px 18px;
    ">
        <b>Legend</b><br>
        <div style="margin-top: 6px;">
            <span style="display:inline-block;width:16px;height:16px;background:blue;border-ra-
dius:50%;margin-right:8px;border:1px solid #333;vertical-align:middle;"></span>
            Vehicle<br>
            <span style="display:inline-block;width:16px;height:16px;background:green;border-
radius:50%;margin-right:8px;border:1px solid #333;vertical-align:middle;"></span>
            Personnel<br>
            <span style="display:inline-block;width:16px;height:16px;background:orange;border-
radius:50%;margin-right:8px;border:1px solid #333;vertical-align:middle;"></span>
            Sensor<br>
            <span style="display:inline-block;width:16px;height:16px;background:purple;border-
radius:50%;margin-right:8px;border:1px solid #333;vertical-align:middle;"></span>
```

**Fig. 7** (continued)

```
        Drone<br>
         <span style="display:inline-block;width:16px;height:16px;background:red;border-ra-
dius:50%;margin-right:8px;border:1px solid #333;vertical-align:middle;"></span>
        Malfunctioning<br>
        <span style="display:inline-block;width:16px;height:16px;background:yellow;border-
radius:50%;margin-right:8px;border:1px solid #333;vertical-align:middle;"></span>
        Charging<br>
          <span style="display:inline-block;width:16px;height:16px;background:cyan;border-
radius:50%;margin-right:8px;border:1px solid #333;vertical-align:middle;"></span>
        In Transit
      </div>
    </div>
    '''
    m.get_root().html.add_child(folium.Element(legend_html))

    return m

# Main logic to simulate asset tracking and generate the map
if __name__ == "__main__":
    NUM_ASSETS = 10  # Number of assets to simulate
    NUM_STEPS = 10  # Number of simulation steps

    print("Simulating asset tracking...")  #simulation is starting
    assets = generate_assets(NUM_ASSETS)
    for _ in range(NUM_STEPS):
        simulate_step(assets)

    final_map = generate_map(assets)
    # Save the generated map to an HTML file
    final_map.save("digital_twin_asset_tracking_map.html")

    # Export asset data to a CSV file
    asset_data = []
    for asset in assets:
        for record in asset['history']:
            asset_data.append({
                "id": asset['id'],
                "type": asset['type'],
                "status": record['status'],
                "latitude": record['coordinates'][0],
                "longitude": record['coordinates'][1],
                "timestamp": record['timestamp']
            })
    df = pd.DataFrame(asset_data)
    df.to_csv("asset_data.csv", index=False)
    print("Asset data exported to asset_data.csv")
    print("Map saved as digital_twin_asset_tracking_map.html")
End
```

**Fig. 7** (continued)

Fig. 8 Simulation image 1

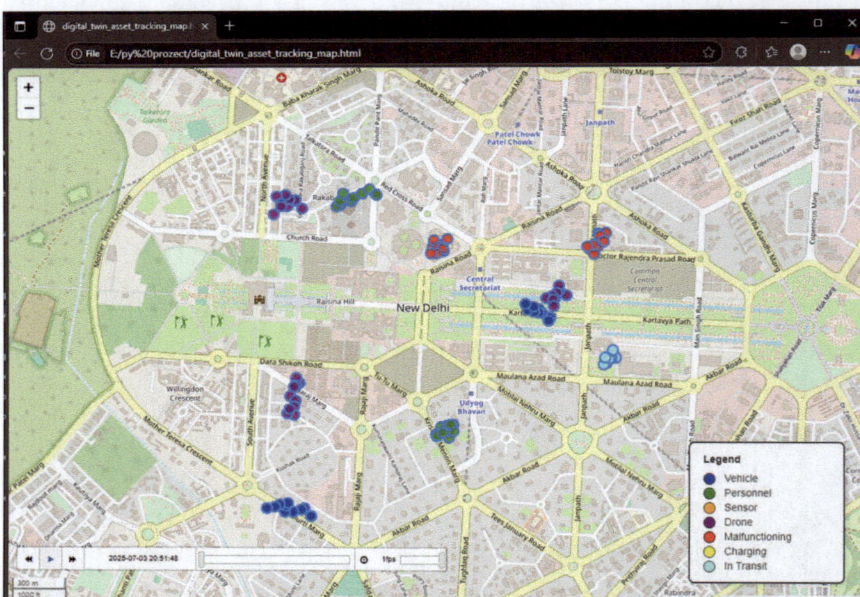


Fig. 9 Simulation image 2

advanced analytics engines provide the foundation for scalable, resilient, and responsive systems that can support diverse sectors, from transportation and utilities to environmental monitoring.

Moreover, this chapter has emphasized the need for robust architectural principles, interoperability standards, and user-focused design to ensure the successful deployment of digital twins in practice. Case studies in logistics, railways, port management, and environmental applications highlight the transformative potential of these systems to improve asset lifecycle management—from planning and construction to maintenance and decommissioning. As the technology continues to mature, trends such as autonomous digital twins, AI-powered decision-making, and the commercialization of Digital Twin-as-a-Service platforms will further enhance their utility. Addressing challenges like data integration, security, and governance will be critical for realizing their promise. Ultimately, geospatially enabled digital twins hold significant potential to create smarter, more resilient, and sustainable infrastructure systems capable of meeting the complex demands of modern society.

## References


1. Kumar R, Singh BA, Sharma M (2025) Understanding data framework for digital twinning. In: Digital twins and simulation technology: Concepts and applications. CRC Press, Taylor & Francis, USA. https://doi.org/10.1201/9781003582489-3
2. Duan H, Gao S, Yang X, Li Y (2025) The development of a digital twin concept system. Digital Twin 2. https://doi.org/10.12688/digitaltwin.17599.3
3. Digital Twin Consortium (2021) Definition and taxonomy of digital twins. Digital Twin Consortium. https://www.digitaltwinconsortium.org
4. International Organization for Standardization (2014) ISO 19115:2014—Geographic information—metadata. ISO, Geneva
5. Open Geospatial Consortium (2016) OGC sensor things API Part 1: Sensing. Open Geospatial Consortium Standard. http://docs.opengeospatial.org
6. Grieves M, Vickers J (2017) Digital twin: Mitigating unpredictable, undesirable emergent behavior in complex systems. Transdisciplinary Perspectives on Complex Systems. Springer, Cham, pp 85–113
7. Kritzinger S, Karner M, Traar G, Henjes J, Sihn W (2018) Digital twin in manufacturing: A categorical literature review and classification. IFAC-PapersOnLine 51(11):1016–1022
8. Batty M (2018) Digital twins. Environ Plan B: Urban Anal City Sci 45(5):817–820
9. Jones A et al (2017) CityGML—Interoperable semantic 3D city models. ISPRS J Photogramm Remote Sens 132:12–33
10. buildingSMART International (2020) Industry Foundation Classes (IFC). buildingSMART International. https://www.buildingsmart.org
11. Goodchild M (2005) GIS and disasters: Planning and response. Ann GIS 11(1):3–20
12. Zerger A, Smith DI (2003) Impediments to using GIS for real-time disaster decision support. Comput Environ Urban Syst 27(2):123–141
13. Schrotter G, Hürzeler C (2020) The digital twin of the city of Zurich for urban planning. PFG–J Photogramm, Remote Sens Geoinformation Sci 88:99–112
14. Madni AM, Madni C, Lucero S (2019) Leveraging digital twin technology in model-based systems engineering. Systems 7(1):7
15. Chapman AC et al (2021) Cyber-physical systems security for smart grid future. IEEE Access 9:29634–29647
16. Colombo A, Karnouskos S, Bangemann T (2014) Industrial cloud-based cyber-physical systems: The IMC-AESOP approach. Springer, Cham
17. Bilal R, Baig AA, Khalid Z (2022) A survey on digital twin: Definitions, characteristics, applications, and design challenges. IEEE Access 10:141422–141449

18. Tao F et al (2020) Digital twin-driven smart manufacturing: Connotation, reference model, applications and research issues. Robot Comput-Integr Manuf 61
19. Cesium (2024) CesiumJS: A javascript library for world-class 3D globes and maps. Cesium. https://cesium.com/cesiumjs
20. Sharma L, Garg PK (eds) (2025) Digital twins and simulation technology concepts and applications, 1st ed. Chapman and Hall/CRC. https://doi.org/10.1201/9781003582489
21. Sharma L, Garg PK (eds) (2024) Deep learning in internet of things for next generation healthcare, 1st ed. Chapman and Hall/CRC. https://doi.org/10.1201/9781003451846
22. Sharma L, Garg PK (eds) (2023) Technological prospects and social applications of society 5.0, 1st ed. Chapman and Hall/CRC. https://doi.org/10.1201/9781003324720
23. Sharma L, Carpenter M (eds) (2022) Computer vision and internet of things: Technologies and applications, 1st ed. Chapman and Hall/CRC. https://doi.org/10.1201/9781003244165
24. Sharma L, Garg PK (eds) (2021) Artificial intelligence: Technologies, applications, and challenges, 1st ed. Chapman and Hall/CRC. https://doi.org/10.1201/9781003140351
25. Sharma L (ed) (2020) Towards smart world: Homes to cities using internet of things, 1st ed. Chapman and Hall/CRC. https://doi.org/10.1201/9781003056751
26. Sharma L, Garg PK (eds) (2019) From visual surveillance to internet of things: Technology and applications, 1st ed. Chapman and Hall/CRC. https://doi.org/10.1201/9780429297922
27. Sharma L (2019) Object detection with background subtraction. LAP LAMBERT Academic Publishing, SIA OmniScriptum Publishing Brivibas gatve 197, Latvia, European Union. (ISBN: 978–613–7–34386–9)

# Chapter 9
# A Proposed Framework for Object Detection in Digital Twin-Driven Geospatial Systems

**Lavanya Sharma**

**Abstract** Digital Twins (DTs) is an emerging technology which revolutionized geospatial system by providing real-time virtual representation of the physical system. The synchronization of digital twins with a physical world is an open challenge, mainly in the detection and tracking of real-time objects. In terms of geospatial system, this technology provides infrastructure management, urban monitoring and environmental analysis with improved spatial intelligence. To maintain the accuracy of DT, dynamic detection of objects from remote sensing data (aerial and satellite images) is very crucial. This chapter presents an artificial intelligence-based object detection using UAV, LiDAR and multispectral data. The proposed work shows high consistency across all object classes such as trees, roads, construction, vehicles and buildings. The performance analysis shows that building and vehicles have higher F1-measure as compared to others. The IoU metric shows that bounding box alignment is accurate and the average of mAP of all 5 objects is 87.3% which validates that the model performs well in urban images while the model performs average in case of construction areas may due to complex background. Further, the advantages, disadvantages, and future work are discussed.

**Keywords** Digital twin · Object detection · Object tracking · Geospatial system · Artificial intelligence · IoU · Accuracy · Remote sensing · GIS · $F1$ measure · Environment · UAV · LiDAR · Multispectral data · Urban monitoring

## 1 Introduction

Digital twins is a virtual replica of physical world objects which exist and play an important role in parallel with their physical form. This technology provides real-time updates of virtual copy to make it act just like real-world object represents.

---

L. Sharma (✉)
Department of IT, New Delhi Institute of Management, Delhi, India
e-mail: shm.lavanya1@gmail.com; shm.lavanya@gmail.com

 
L. Sharma et al. (eds.), *Digital Twin Technology in Geospatial Science: Trends, Applications and Challenges*, Data-Intensive Research,
https://doi.org/10.1007/978-3-032-20311-3_9

It is combination of various cutting-edge technologies such as Internet of Things, cloud computing, geographic information systems (GIS), artificial intelligence and machine learning. Among these technologies GIS provides location, data for geospatial digital twins and enables its mapping. In this AI plays a vital role; it processes sensor data and detects trends and pattern for modeling, simulations and predictive analysis. IoT provides the required sensor or camera data to create and update the virtual copy. Here, visualization enables creations and 3D replica of real-time objects. Geospatial systems form the backbone of various real-time applications such as smart cities, environmental monitoring, visual surveillance and disaster management. Nowadays, there is rapid growth of urban complexity, so the need for real-time representation of geospatial environment has become very crucial.

Digital twins has provided a foundational change that allows two-way communications between real and virtual world. The synchronization of digital twins with a physical world is an open challenge, mainly in the detection and tracking of real-time objects. Several research studies are in progress to handle object detection and digital twins, but few have evaluated its significance within geospatial systems. The conventional object detections approach in remote sensing used the statistical classifier. With the rise of advanced techniques such as deep learning models such as Yolo, RetinaNet and Faster R-CNN (convocational neural network) shows a significant improvement in the detection of motion and occluded objects in satellite imagery. Recent studies have extended the above-listed models to geospatial datasets such as SpaceNet, xView, and DOTA that show outstanding outcomes in terms of speed and accuracy. These models are used to update digital twins dynamically. In urban systems, digital twin's implementation relies on sensor data, building information modeling (BIM), and simulation-based inputs [1–18].

The detection of motion-based object is a sub-domain of computer vision that detects and localizes the object in video data or digital images. Object detection in GSS involves detecting building, roads, vegetation, moving vehicles and animals captured from satellite imagery and drone imagery. Although there is a technological advancement in this domain, the mapping of object detection with digital twin architecture has yet to be thoroughly investigated. This chapter proposes a framework that combines state-of-the-art object detection techniques with digital twin's architecture. The traditional geospatial system relies on manual updates and survey, but this proposed framework utilizes AI-based object detection timely environment validation. This combined technique can support predictive modeling and also help in real-time decision-making in various domains. The main steps are: (i) end-to-end architecture for object detection-based digital twin, (ii) evaluation of framework using real-time datasets and other enhancement tools and (iii) performance analysis using various parameters such as precision, recall and latency in the urban environment.

## 2 Related Work

Recent advancement in smart cities highlights the importance of spatial intelligence in urban development. Various initiatives such as CityGML and UrbanSim offer valuable tools for urban planning and management. These tools do not provide autonomous updates based on real-time scenarios. Some recent work has shown improvement by combining computer vision and GIS but often fails to coordinate with the physical environment. In recent years, the digital twins with geospatial intelligence has shown a tremendous growth. Grieves and Vickers [11] present a real-time-based digital twins model for virtual representation of the physical world. Batty [3] further extends this to the urban system by providing digital twins as a tool for simulation and managing city infrastructure.

At the same time, object detection models have also shown a significant rise that enables real-time visual surveillance in spatial systems. Redmon and Farhadi's [12] highlights YOLOv3 for fast detection of objects and Liu et al. introduced "Swin Transformer" more suited for dense spatial layouts. The literature shows a clear transition from conventional computer vision and motion detection techniques to advanced intelligent systems. In another work, Anirban Dutta et al. (2020) focuses on advanced object detection tracking techniques, emphasizing deep learning-based enhancements. Nilanjan Dey et al. (2016, 2018) focus on feature detection, motion analysis, and the importance of big data in remote sensing for efficient analysis and visualization. Recently, work done by Md Rafiqur Tonmoy et al. (2025) introduces a new hybrid Vision Transformer models for image classification. In this paper, machine learning is also applied and smart city frameworks further demonstrate the combination of intelligent systems for real-world applications. Sharma et al. [13, 17], Sharma and Garg [15], Sharma and Carpenter [14], Yadav [20], Baral et al. [2] have presented an efficient motion-based object technique in background modelling (Background Subtraction methods),comparative analysis and literature review, which overall led to enhanced video surveillance methods. In another work, Dutta et al. [10, 6] and Dey et al. [4] have contributed significantly to this domain of tracking and vision. Agarwal et al. [1] proposed a GIS and remote sensing-based framework to detect suitable artificial groundwater recharge sites using spatial analysis and hydrogeological parameters. Sharma et al. [16] presented a Fisher's linear discriminant (FLD) ratio-based thresholding technique for accurate object detection in thermal videos. Furthermore, more advancements in object detection such as Kaiwen Duan et al. [9] with CenterNet and Piotr Dollár et al. [8] with feature pyramids significantly improved detection efficiency, while Shanshan Zhang et al. [21] analyzed the challenges and progress in pedestrian detection.

# 3 Proposed Framework

This section presents a framework and its interconnected modules. Apart from the recent advances, several studies offer a unified system for real-time object detection with dynamically updating Geospatial digital twins. The proposed framework handles this gap by providing a combined approach of deep learning, remote sensing-based data collection and spatial intelligence to build a self-updating and decision support system for urban planning as shown in Fig. 1. The proposed system consists of the listed steps:

1. Remote sensing (input layer) and pre-processing (Georeferencing).
2. AI-based object detection.
3. Digital twins synchronization engine.
4. Geospatial analytics and visualization interface.

Each step is built to be modular and scalable which allows deployment of cloud and edge environment. This architecture supports batch and real-time processing which makes it more suitable for continuous visualization of large geospatial zones or areas as shown in Fig. 2.

## 3.1 *Remote Sensing and Pre-processing*

Data is collected from multi-spectral satellite sensors such as Sentinel-2, WorldView, high-resolution UAVs (Unmanned Aerial Vehicles) and LiDAR (Light Detection and Ranging). Pre-processing includes normalization, geo-rectification (image alignment), shadow removal and image segmentation. In this step, image tiles are created using sliding window to make spatial coverage while reducing computational load. The georeferencing can be done using metadata and GCP (ground control points), that allow alignment with GIS layers. Each tile has temporal and spatial references for historical comparison.

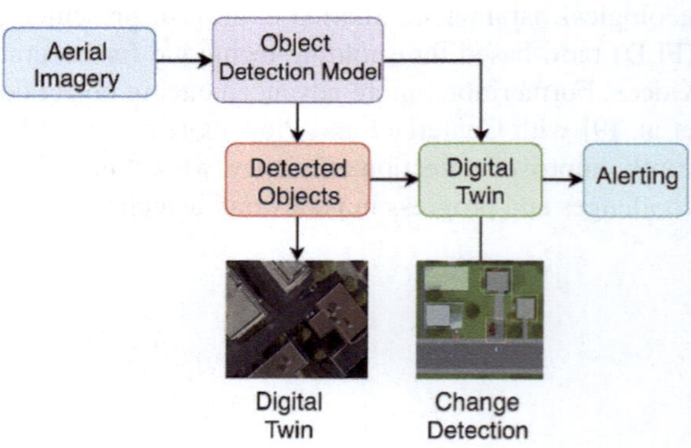


**Fig. 1** Object detection (aerial imagery) into a digital twin for change detection and alerting

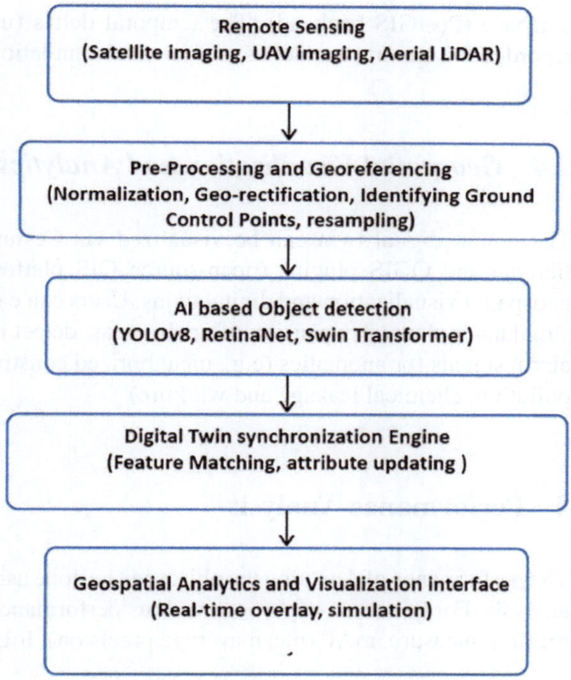


**Fig. 2** Proposed framework for object detection in digital twin-driven geospatial systems

## 3.2 *AI-Based Object Detection*

This proposed model consists of 2 deep learning models: (i) YOLOv8 which is used for real-time detection and (ii) Swin Transformer which is used for high accuracy environment. The model can be trained on a hybrid dataset comprising of buildings, roads, vehicles, trees, aircraft and many more. The models are evaluated using performance metrics like Precision, Recall, F1-score and mAP (mean average precision).

- SpaceNet (e.g. buildings, roads, streets).
- DOTA (vehicles and aircraft).
- dataset (urban utilities, trees, plants and pollution sites).

## 3.3 *Digital Twin Synchronizer*

The detected objects can be mapped onto a 3D GIS-enabled digital twin using spatial interpolation (technique used in geospatial analysis) and feature matching (detects key points such as building corners, road intersections, or tree clusters). This module updates entity attributes such as dimensions, location and state in the digital twin database (PostGIS backend). The temporal deltas (used in change detection) are recorded to support time-series analysis and simulations.

### 3.4 Geospatial Visualization and Analytics

The output digital twin can be visualized via CesiumJS (open-source JavaScript library) and QGIS plugins (open-source GIS platform). These both support 3D geospatial visualization and digital twins. Users can easily check real-time overlays, simulate unplanned expansion of urban areas, detect motion of object and generate alarm signals for anomalies (e.g., unauthorized constructions, flood, earthquake, air pollution, chemical leakage and wild fire).

## 4 Performance Analysis

The performance of detecting the object can be done using qualitative and quantitative analysis. For quantitative purpose various performances are used such as precision, recall, $f$-measure, mAP (mean average precision), IoU (Intersection over union).

### 4.1 Qualitative Analysis

The qualitative analysis is done by comparing the detected bounding boxes with the original aerial image. The geospatial placement is verified using the CesiumJS-based digital twin (Fig. 3).

A spatial alignment test shows 88% of the detected object has an offset of <1.5 m from its ground truth which makes it suitable for the GIS layer. The below detected image represents YOLOv8 detection overlaid with bounding boxes: (i) buildings (blue), (ii) vehicles (red), trees (green), construction areas (orange) (Fig. 4).

### 4.2 Quantitative Analysis

To measure the effectiveness of the detected object the evaluation can be done using the various performance metrics (Precision, Recall, $F1$-Score, IoU, and mAP@0.5 across five object classes) on UAV images that contain urban, residential and constructed environment as shown in Table 1.

$$\text{Precision} = \text{TP} + \text{FP}/\text{TP} \tag{1}$$

**Fig. 3** Original aerial image (UAV)

**Fig. 4** Object detection (overlay) with bounded boxes

**Table 1** Object detection outcome per class

| Object class | Precision | Recall | $F1$-score | IoU | mAP@0.5 |
|---|---|---|---|---|---|
| Buildings | 93.5 | 90.1 | 91.7 | 0.85 | 91.2 |
| Vehicles (cars) | 88.4 | 86.9 | 87.6 | 0.80 | 88.1 |
| Trees | 90.2 | 85.6 | 87.8 | 0.81 | 87.5 |
| Roads | 89.2 | 83.8 | 86.5 | 0.79 | 85.3 |
| Construction | 87.0 | 81.5 | 84.2 | 0.77 | 83.9 |
| Average | 89.6 | 85.6 | 87.5 | 0.80 | 87.2 |

$$\text{Recall} = \text{TP} + \text{FN}/\text{TP} \tag{2}$$

$$F1\text{-measure} = 2 \cdot \text{Precision}^{*}\ \text{Recall}/\text{Precision} + \text{Recall} \tag{3}$$

$$\text{IoU} = \text{Area of Union}/\text{Area of Overlap} \tag{4}$$

The proposed work shows high consistency across all object classes (trees, roads, construction, vehicles, buildings). As shown in Table1 building and vehicles have higher $F1$-measure as compared to others. The IoU metric shows that bounding box alignment is accurate and the average of mAP of all 5 objects is 87.3% which validates that the model performs well in urban images while the model performs average in case of construction areas which may be due to a complex background.

### *4.3 Digital Twin Synchronization and Urban Intelligence*

The next step is digital twins synchronization; in this step the detected object can be stored in a spatial database (PostGIS) with CesiumJS visualization. Object tracking and change detection can be performed using spatial temporal queries by comparing the current and past image data.

## 5 Discussion

The proposed framework performs well in a urban geospatial environment. The main advantage of this work is an automated, self-updating loop which minimizes the human intervention. Additionally, LiDAR enhanced the detection in dense urban zones. This object detection-based digital twin framework has one limitation which is its dependency on image quality and illumination conditions especially in case of a cloudy environment, occluded objects. The present implementation is also limited due to its limited resources. The privacy concern must be addressed by data governance policies, where energy consumption of AI models needs optimization for sustainability is one of the main constraint. Despite all these limitations and challenges, the framework presents an innovative approach towards completely autonomous geospatial digital twins. Future work includes semantic segmentation, 3D models for simulation, temporal tracking (deals with dynamic movement), and privacy compliance.

## 6 Conclusion

This chapter presents a framework that combines object detection techniques with digital twin's driven geospatial system. The conventional geospatial system relies on manual updates and survey, but this proposed framework utilizes AI-based object detection to timely validate the environment. This combined technique can support predictive modelling and also help in real-time decision-making in various domains. The analysis is done on UAV images. The system has high spatial accuracy and detection rate across various urban object classes (buildings, trees, vehicles). The automated, self-updating loop reduces the human intervention while maintaining accuracy with ground truth. Apart from open challenges related to illuminations, occlusion, complex background and energy constraints, the proposed framework marks a significant step towards dynamic, urban systems. This object detection-based digital twin framework has one limitation that its dependency on image quality and illumination conditions especially in case of cloudy environment, occluded objects. In future we will explore semantic segmentation, 3D models for simulation and temporal tracking.

## References


1. Agarwal R, Garg PK, Garg RD (2013) Remote sensing and GIS based approach for identification of artificial recharge sites. Water Resour Manag 27:2671–2689. https://doi.org/10.1007/s11269-013-0310-7
2. Baral A, Gupta D, Sharma L (2019) Motion based object detection based on background subtraction: a review. In: 3rd international conference on electronics, communication and aerospace technology (ICECA 2019). IEEE, Coimbatore, pp 545–549. https://doi.org/10.1109/ICECA.2019.8822128
3. Batty M (2018) Digital twins. Environ Plan B Urban Anal City Sci 45(5):817–820. https://doi.org/10.1177/2399808318796416
4. Dey N, Ashour A, Patra K (eds) (2016) Feature detectors and motion detection in video processing. IGI Global
5. Dey N, Bhatt C, Ashour AS (2018) Big data for remote sensing: visualization, analysis and interpretation. Springer, Cham, p 104
6. Dey N, Bhatt C, Ashour AS (2018) Big data for remote sensing: visualization, analysis and interpretation. Springer, Cham, p 104

7. Dey N, Wagh S, Mahalle PN, Pathan MS (eds) (2019) Applied machine learning for smart data analysis. CRC Press
8. Dollár P, Appel R, Belongie R, Perona P: Fast feature pyramids for object detection. IEEE Trans Pattern Anal Mach Intell 36(8):1532–1545 (2014)
9. Duan K, Bai S, Xie L, Qi H, Huang Q, Tian Q: Centernet: keypoint triplets for object detection. In: Proceedings of the IEEE international conference on computer vision, pp 6569–6578 (2019)
10. Dutta A, Mondal A, Dey N, Sen S, Moraru L, Hassanien AE (2020) Vision tracking: a survey of the state-of-the-art. SN Comput Sci 1(1):57
11. Grieves M, Vickers J (2017) Digital twin: mitigating unpredictable, undesirable emergent behavior in complex systems. In: Transdisciplinary perspectives on complex systems. Springer, pp 85–113. https://doi.org/10.1007/978-3-319-38756-7_4
12. Redmon J, Farhadi A (2018) YOLOv3: an incremental improvement. https://arxiv.org/abs/1804.02767
13. Sharma L et al (2025) An introduction to digital twins and simulation. In: Digital twins and simulation technology: concepts and applications, vol 1. Taylor and Francis CRC Press, pp 1–12
14. Sharma L, Carpenter M (eds) (2022) Computer vision and internet of things: technologies and applications, 1st edn. Chapman and Hall/CRC. https://doi.org/10.1201/9781003244165
15. Sharma L, Garg PK (eds) (2025) Digital twins and simulation technology: concepts and applications (1st ed). Chapman and Hall/CRC. https://doi.org/10.1201/9781003582489
16. Sharma L, Yadav DK, Singh A (2016) Fisher's linear discriminant ratio based threshold for moving human detection in thermal video. Infrared Phys Technol 78:118–128. ISSN 1350-4495. https://doi.org/10.1016/j.infrared.2016.07.012
17. Sharma L, Yadav DK, Bharti SK (2015) An improved method for visual surveillance using background subtraction technique. In: 2nd international conference on signal processing and integrated networks (SPIN), Noida, India, pp 421–426. https://doi.org/10.1109/SPIN.2015.7095253
18. Solanki VK, Katiyar S, Bhashkar Semwal V, Dewan P, Venkatasen M, Dey N (2016) Advanced automated module for smart and secure city. Procedia Comput Sci 78:367–374
19. Tonmoy MR, Hossain MM, Dey N, Mridha MF (2025) MobilePlantViT: a mobile-friendly hybrid ViT for generalized plant disease image classification. arXiv:2503.16628
20. Yadav DK (2016) Histogram based adaptive learning rate for background modelling and moving object detection in video surveillance. Int J Telemed Clin Pract 2(1):74–92. https://doi.org/10.1504/IJTMCP.2017.082107
21. Zhang S, Benenson R, Omran M, Hosang J, Schiele B: How far are we from solving pedestrian detection? In: Proceedings of the IEEE conference on computer vision and pattern recognition, pp 1259–1267 (2016)
22. Zhou Y, Tuzel O (2018) VoxelNet: end-to-end learning for point cloud based 3D object detection. In: CVPR, pp 4490–4499. https://doi.org/10.1109/CVPR.2018.00472

# Chapter 10
# Hypergraph Convolutional Refinement in Latent Space for Underwater Digital Twins

**Meghna Kapoor, Badri Narayan Subudhi, Thierry Bouwmans, and Ankur Bansal**

**Abstract** Detecting objects in underwater scenes is essential for constructing Digital Twin (DT) models tailored to marine settings. These models enable applications like monitoring underwater infrastructure, evaluating ecological conditions, and guiding autonomous underwater systems. However, accurate detection remains challenging due to complex optical distortions and limited visibility inherent in underwater scenes. Existing deep learning methods often struggle to capture the fine-grained contextual relationships required for reliable representation within DT frameworks. In this work, we propose a novel end-to-end architecture that integrates hypergraph based latent space refinement to improve underwater object detection for Digital Twin applications. Our approach begins with a convolutional encoder that projects raw underwater imagery into a latent feature space, where an unsupervised graph is constructed. Leveraging hypergraph convolution, the model enables robust message passing between multiple latent variables simultaneously, enhancing the representation of spatial and semantic relationships. The refined graph is then decoded back into the image space to recover object-level information. We validate the effectiveness of our method through extensive experiments on benchmark underwater datasets, demonstrating superior performance over fourteen state-of-the-art models in preserving object boundaries and extracting critical detail for high-fidelity Digital Twin construction.

M. Kapoor · B. N. Subudhi · A. Bansal
Indian Institution of Technology, Jammu, Jammu and Kashmir, India
e-mail: meghna@iitjammu.ac.in

A. Bansal
e-mail: ankur.bansal@iitjammu.ac.in

T. Bouwmans (✉)
Laboratoire MIA, La Rochelle Universite, La Rochelle, France
e-mail: thierry.bouwmans@univ-lr.fr

 
L. Sharma et al. (eds.), *Digital Twin Technology in Geospatial Science: Trends, Applications and Challenges*, Data-Intensive Research,
https://doi.org/10.1007/978-3-032-20311-3_10

# 1 Introduction

To enable intelligent monitoring and environmental understanding in marine ecosystems, Digital Twin (DT) systems for underwater environments are gaining increasing attention. A key component of such systems is the ability to accurately perceive and interpret visual data acquired in situ. However, underwater imaging presents unique challenges due to light absorption, scattering, and wavelength-dependent attenuation. These phenomena result in color distortion, low contrast, and texture loss, which significantly degrade image quality. Such degradation limits the effectiveness of vision-based models, especially in detecting and tracking objects, which are critical tasks for constructing reliable underwater digital replicas.

Deep learning has transformed visual monitoring systems, offering enhanced object recognition even in environments with high activity levels, such as urban areas and commercial hubs. Traditional surveillance methods [1] typically model objects as either foreground or background. However, this approach becomes increasingly challenging when objects change shape and structure across frames. Such variability complicates the accurate learning and prediction of moving object structures. These challenges are even more pronounced in underwater environments, where color and texture cues crucial for object modeling are significantly degraded due to light absorption [2].

Underwater moving object detection has been sparsely addressed in comparison to the extensive work done in terrestrial surveillance. Traditional methods like Gaussian Mixture Models (GMM) [3] and Kernel Density Estimation (KDE) [4] perform well under the assumption of minimal environmental noise. However, these methods struggle to detect objects effectively in underwater environments, where color cues are often unreliable. Approaches such as those proposed in [5, 6] exploit texture features to differentiate objects from the background, but factors like waves and the movement of seaweed and algae introduce significant noise that complicates detection. Rout et al. [7] proposed a method that integrates the Wronskian framework into GMM to better model underwater noise. However, the said method can still not procure satisfactory results due to complications present in the underwater environment. Conventional visual cues that work in conventional surveillance scenarios become unreliable in underwater. Additionally, objects in underwater settings often display erratic, three-dimensional movements, making motion and structurebased features less effective. Recent studies [8–10] have highlighted the potential of deep learning approaches, particularly encoder-decoder architectures, in extracting abstract features that reliably represent objects. However, these methods often struggle to preserve intricate object details without significant information loss in challenging environments. Giraldo et al. [11] demonstrated that graph-based representations, derived from features extracted through various methods, can effectively represent objects. Yet, these handcrafted features tend to degrade, making it difficult to distinguish objects from the background.

To address this issue, Kapoor et al. [12] recently demonstrated that applying graph theory in latent space can significantly mitigate these effects and preserve more information. Despite this advancement, their approach relied on a basic graph structure, which is insufficient for capturing the complex relationships among latent variables. Therefore, adopting a more sophisticated graph structure could enhance object detection accuracy in such challenging environments. The conventional CNN architectures retain noise due to the non-Euclidean nature of the convolution kernels in the convolution space after multiple processing layers. This issue disrupts the spatial relationships within images, thereby introducing noise into the feature abstraction. Moreover, while pooling layers are integral to ensuring translation invariance within convolutional kernels [13], which inadvertently introduce noise that cannot be fully mitigated during the re-projection into image space, leading to the permanent loss of fine boundary information. However, Kapoor et al. [12] showed that applying graph theory techniques can significantly reduce noise, leading to more accurate boundary delineation and improved object detection in complex environments. Nevertheless, the relationships between latent variables are often not straightforward; an edge might connect two or more vertices. A more sophisticated approach should involve initializing an edge with connections to multiple vertices to capture these complex relationships better. Furthermore, determining the connections among graph nodes becomes particularly challenging in scenarios where no ground truth is available, making it difficult to establish accurate relationships between the nodes.

In this paper, we propose a novel underwater object detection technique by combining the spatial relationships inherent in convolutional networks (CNN) with the structural relationships of graph refactoring. We employ a ResNet-50 encoder for robust feature extraction and efficient projection of input image frame into the latent feature space. The feature vector encapsulates the attributes of the object, and by sharing information among latent variables, we ensure the preservation of essential details. The proposed method enhances the retention of spatio-contextual information by initializing an unsupervised graph using the latent vector, resulting in a more comprehensive representation of spatial relationships. To refine the complex relationship of the latent vector, we introduce a hypergraph which models the relationship among two or more latent variables as a hyperedge for efficient message passing and retention of intricate details. Additionally, we address the challenge of graph initialization in scenarios where node labels are unknown. Hypergraph convolution is employed to enhance information retention among the graph nodes, hence the object boundaries in underwater. Our architecture integrates an innovative image generation module that reconstructs the image, with weight sharing between the CNN backbone and the image generation module. This ensures the preservation of minute details and semantic features, maintaining fine-grained information throughout the detection process and enhancing the overall accuracy and reliability of underwater object detection.

The main contributions of this work are as follows:

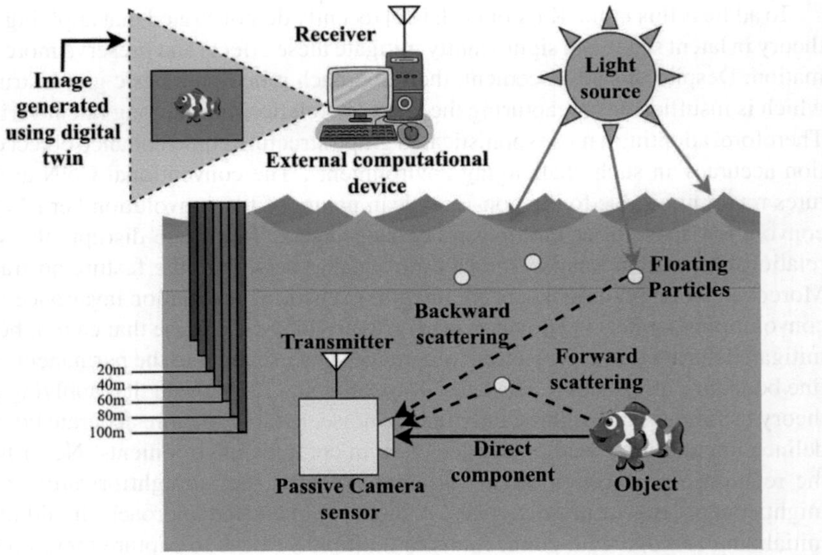


**Fig. 1** An example of using digital twin technology to create a virtual twin of underwater object using a passive camera sensor

- We present a novel end-to-end deep learning architecture that integrates graph refactoring into an encoder-decoder framework. This design is specifically developed for detecting moving objects in challenging underwater environments and is suitable for use in Digital Twin systems for marine applications.
- To capture complex dependencies among latent features, we introduce a hypergraph based projection approach. This allows for more accurate modelling of interactions between multiple latent variables, improving the representation of fine structural details.
- We develop an unsupervised method for graph initialization that does not rely on labeled node data. This is particularly useful in underwater scenarios where annotated datasets are limited or unavailable.
- We perform a comparative analysis of message passing strategies in latent hyperspace. Specifically, we evaluate convolutional and attention-based kernels to determine which approach more effectively preserves semantic information and object boundaries (Fig. 1).

## 2 Research Background

Moving object detection in underwater environments is a nontrivial task due to significant visual degradation caused by light absorption, scattering, suspended particles, and dynamic background interference. Although numerous methods have been proposed for object detection in terrestrial and aerial contexts, relatively few have been adapted to address the specific challenges presented by underwater scenarios. The existing body of literature may be grouped into three primary categories: conventional techniques, learning-based approaches, and graph-based models.

## 2.1 Conventional Methods

Traditional approaches often treat each pixel as belonging either to the foreground or background, relying primarily on motion or appearance cues. In underwater environments, the motion of elements such as algae or seaweed can produce quasi-periodic changes that are misclassified as object movement [10, 21, 22, 25, 38, 48, 48, 50]. Techniques based on frame differencing and background subtraction have been applied, but their performance tends to degrade in the presence of dynamic backgrounds. For instance, Rout et al. [14] introduced a five-frame difference model to detect motion, while Prabowo et al. [15] constructed a background model and subtracted incoming frames to identify foreground regions. These approaches, although effective in simple scenes, are vulnerable to noise from water motion and aquatic vegetation. Rout et al. [7] further extended the background modeling using a Wronskian-based Gaussian Mixture Model (GMM), aiming to better handle background complexity. Pagire and Phadke [16] applied a mixture of Gaussians to improve robustness in noisy underwater conditions. Probabilistic methods have also been explored. Salman et al. [17] modeled pixel values using posterior distributions for background estimation. However, these methods often fail when objects blend into the background or when occlusion is present. To address issues caused by color distortion, Vasamsetti et al. [6] used both color and texture features to distinguish objects from their surroundings. Later, Vasamsetti et al. [5] proposed a method based on three-dimensional local spatio-temporal ternary patterns, suggesting that color alone may not be sufficient for underwater detection. Despite these improvements, many conventional models lack the ability to generalize in scenes with low contrast or strong visual degradation.

## 2.2 Learning Based Methods

Deep learning techniques have significantly improved the capacity to detect and segment objects in complex scenes by eliminating the need for handcrafted features. These models can learn discriminative patterns directly from raw image data. Fu et al. [18] employed Mask R-CNN to detect underwater objects as regions of interest, while Song et al. [19] enhanced the network by modifying the module to reduce errors in object probability estimation. Although these models achieve high accuracy in many scenarios, they often underperform in preserving fine object boundaries, especially when color and texture information is degraded. Encoder-decoder architectures have also been applied to underwater detection. Bajpai et al. [8] introduced a U-Net-based model to retain structural details during object segmentation. Similarly, Tseng and Lin [20] proposed a UNet++ model that incorporates data augmentation to compensate for limited training data. These architectures have demonstrated the ability to maintain shape and contour, though they may still struggle in noisy environments where semantic consistency is difficult to capture. Generative Adversarial Networks (GANs) offer another approach by learning to differentiate between real and generated image data. Patil et al. [10] proposed a GAN-based framework using motion saliency to detect moving underwater objects. Although GAN scan adapt to novel data distributions, their training process is typically unstable and demands extensive datasets to achieve consistent performance. More recently, attention mechanisms and transformer-based architectures have been introduced for underwater object detection. Sun et al. [21] utilized the Swin Transformer to extract local and global contextual features, achieving performance comparable to CNN-based models. Liu et al. [22] improved this approach by combining shallow and deep feature maps and integrating online learning techniques. Despite their accuracy, transformer models often incur high computational costs and may not exploit spatial relationships as effectively as other architectures.

### *2.3 Graph-Based Methods*

Graph-based models are increasingly recognized for their capacity to handle data with irregular structures. These models operate on non-Euclidean representations and are well suited for complex tasks such as object segmentation and contextual understanding. Graph Convolutional Networks (GCNs), for example, propagate information between connected nodes based on spatial or semantic relationships. Xu et al. [23] developed a spatially aware graph learning method that uses learnable Gaussian kernels for contextual feature extraction. Chen et al. [24] proposed integrating semantic segmentation outputs with graph structures by initializing the adjacency matrix from image-derived features. GCNs offer an effective mechanism for modeling non-uniform spatial relationships and have been used in various vision tasks. In the domain of visual object interaction, Liang et al. [25] constructed a semantic graph using attention mechanisms to identify relationships between objects. He et al. [26] applied a similar concept to generate 3D point clouds by constructing voxel graphs and learning discriminative features. These models are powerful in structured environments but are rarely applied in underwater contexts. Giraldo et al. [11] introduced a semi-supervised graph-based method known as GraphMOS for moving object segmentation in surveillance videos. This framework integrates segmentation, feature extraction, background modeling, graph construction, and optimization into a unified pipeline. Although GraphMOS shows superior performance in terrestrial scenes, it does not account for the distortions and noise present in underwater videos. To date, few graph-based approaches have been adapted for underwater object detection, leaving a gap in the development of robust graph learning models for such settings.

# 3 Proposed Method

The proposed end-to-end architecture, illustrated in Fig. 2, processes input images through three main components: a convolutional backbone for feature extraction, a hypergraph-based latent space module for structural representation, and an image generation module for final object detection. The system is designed to enhance underwater object detection by refining latent representations through a learned hypergraph structure. The refined latent features are then decoded into a segmentation map that highlights moving objects in underwater scenes.

## *3.1 Backbone Network*

The feature extraction stage uses ResNet-50 as the backbone. This network comprises multiple convolutional blocks with ReLU activations and max pooling layers to extract spatial hierarchies from the input image. Four blocks are used to progressively capture increasingly abstract representations. After the final convolutional layer, the network produces a feature map of size $7 \times 7$, which is then flattened into a latent vector of size $1 \times 49$. This latent representation encodes both spatial and semantic information and serves as the foundation for graph construction.

## *3.2 Graph Construction*

A graph is formally defined as an unordered pair G = (V, E), where V is the set of $N$ vertices and E is the set of edges. In the context of image-based learning, node features may represent colour, motion, or texture attributes, as shown by Giraldo et al. [11]. However, in underwater imagery, such features are often unreliable due to severe degradation from scattering, colour cast, and turbidity.

To address this limitation, the present work constructs the graph directly in the latent space produced by the convolutional backbone. Each dimension of the latent vector is interpreted as a node $v_i$ with its associated feature. The construction process involves the following steps:

1. **Latent Nodes**: Each element $z_i$ of the flattened latent vector is treated as a node in the graph, resulting in the vertex set $V = \{v_1, v_2, \ldots, v_N\}$. These nodes represent distinct spatial or semantic features extracted from the input image.

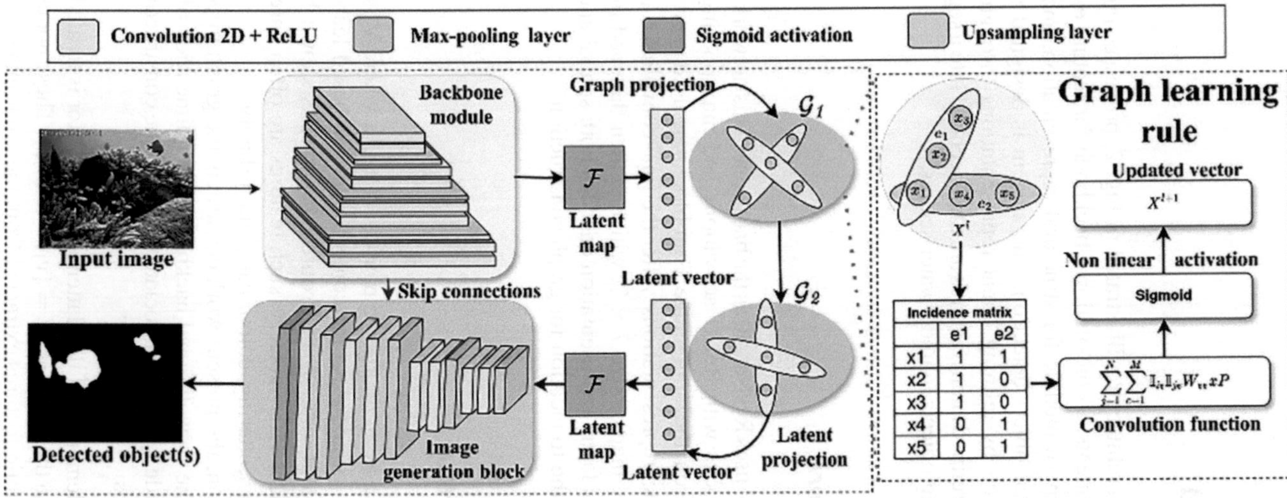


**Fig. 2** The proposed end-to-end architecture comprises three components: a backbone, a graph learning module, and an image generation block. The backbone processes the input image to generate a feature vector representing its abstract features. This vector is refined by the graph learning module to maintain spatio-contextual relationships. The refined vector is then projected into latent space and passed to the image generation module to produce a binary segmented image

2. **Initial Connectivity**: The incidence matrix I, which defines the connectivity between nodes and hyperedges, is initialized by sampling from a random distribution. This structure is progressively refined during training through backpropagation, allowing the network to learn task-specific relationships in the latent space.

The resulting hyper-graph captures both local similarity and global context.

## 3.3 Hypergraph Convolution Module

While simple graphs model pairwise relationships, hypergraphs enable connections among multiple nodes through hyperedges [27]. This is particularly valuable in the latent space, where interactions among groups of features are often more informative than binary relations.

The graph $\mathcal{G} = (\mathcal{V}, \mathcal{E})$ is extended into a hypergraph by defining a set of hyperedges $\mathcal{H}$, where each hyperedge connects more than two nodes. Each node corresponds to an element of the latent vector, and each hyperedge connects a subset of semantically related nodes. The degree of a vertex is defined as:

$$D_v = \sum_{i=1}^{M} W_{ee} \mathbb{I}_{ie}$$

Here, $W_{ee}$ is the weight assigned to hyperedge $e$, and $\mathbb{I}_{ie}$ is the incidence matrix entry, which equals 1 if node i is part of hyperedge e, and 0 otherwise.

The degree of a hyperedge, representing the number of vertices it connects, is given by:

$$D_h = \sum_{i=1}^{N} \mathbb{I}_{ie}$$

Vertices sharing a hyperedge are more strongly connected and participate in collaborative feature refinement. Node features are updated via hypergraph convolution using:

$$x_i^{l+1} = \sigma\left(\sum_{j=1}^{N}\sum_{e=1}^{M} \mathbb{I}_{ie}\mathbb{I}_{je} W_{ee} x_j^l P\right)$$

where $\sigma(\cdot)$ is a nonlinear activation function, $P$ is the layer weight matrix, and $l$ denotes the layer index. This can also be represented in matrix form as:

$$x^{l+1} = \sigma\left(D_v^{\frac{-1}{2}} \mathbb{I} W D_e^{\frac{-1}{2}} \mathbb{I}^T D_v^{\frac{1}{2}} x^l P\right)$$

To further improve relational modeling, attention mechanisms are used to construct dynamic incidence matrices. Instead of fixed hyperedges, the model learns how strongly each node influences others within a hyperedge, allowing for adaptive information flow.

### 3.4 Image Generation Module

Once node features are refined via hypergraph convolution, they are reshaped into a matrix suitable for decoding. This feature map, enriched with spatial and contextual information, is passed through transposed convolutional layers followed by upsampling to produce the final segmentation output.

To preserve consistency and high-level semantics, weights are shared between the encoder and decoder components. This design ensures that learned representations are coherent across the network. The model is trained using a binary cross-entropy loss applied to the segmentation output, promoting accurate separation of foreground (moving objects) from background.

## 4 Experimental Results

This section presents the experimental setup and performance analysis of the proposed method. Both qualitative and quantitative comparisons are conducted against state-of-the-art techniques. Additionally, an ablation study is performed to investigate the effect of various architectural components and hyperparameters.

### 4.1 Implementation Details

The proposed end-to-end architecture was implemented using PyTorch and trained on an NVIDIA A100 GPU with 80 GB of memory. The model was trained for 100 epochs with a batch size of 8. Stochastic Gradient Descent (SGD) was used as the optimizer with an initial learning rate of $10^{-4}$. A StepLR scheduler was applied to reduce the learning rate at specified intervals to enhance convergence. All code and additional results are available for reproducibility at[1]:

---

[1] https://github.com/immkapoor/hyper_graph.

**Algorithm 1:** Proposed Hypergraph Refactored Method

**Input:** Input image $I$

**Output:** Detected moving objects O

**Feature Extraction via Backbone**;

Extract feature map F from image I using ResNet-50 backbone;

Flatten F into latent vector $x^0$ of size $1 \times 49$;

**Hypergraph Construction and Convolution**;

Initialize nodes $V \leftarrow \{x_1^0, x_2^0, \ldots, x_{49}^0\}$;

Initialize random incidence matrix $I \in R^{N \times M}$;

Initialize hyperedge weights $W_{ee}$ and layer weight matrix $P$;

**for** *each layer* $l = 0$ *to* $L - 1$ **do**

  Compute vertex degree matrix $D_v$;

  Compute hyperedge degree matrix $D_e$;

  Update node features:;

$$x^{l+1} = \sigma\left(D_v^{\frac{-1}{2}}\, \mathbb{I} W D_e^{\frac{-1}{2}}\, \mathbb{I}^T D_v^{\frac{1}{2}}\, x^l P\right)$$

  **if** *using attention* **then**

    Update I based on attention scores among nodes;

**Image Generation and Object Detection**;

Reshape $x^L$ into feature map $F'$;

# 5 Qualitative Analysis

Figure 3 presents a visual comparison of the proposed method against seven state-of-the-art techniques using the Fish4Knowledge dataset. Three key challenges are considered: complex backgrounds, crowded scenes, and dynamic environments. In the first scenario, the object's appearance blends with surrounding elements such as rocks. Traditional methods like MultiCueBGS [28] and SuBSENSEBGS [29] struggle to detect the object when its color closely matches the background. Techniques such as Texture BGS [30] and ML-BGS [31] fail to preserve object shape, while deep learning approaches [8, 32] also have difficulty maintaining structural integrity. The proposed method preserves object contours more effectively, though minor holes may appear due to graph aggregation. In the second scenario involving crowded scenes, conventional methods [28–31] perform poorly. Deep learning-based methods [8, 32] often merge multiple objects into indistinct blobs. In contrast, the proposed model is capable of distinguishing and retaining clear boundaries, even when multiple fish are present. The third challenge involves dynamic backgrounds where objects are partially occluded. Traditional methods frequently misclassify background motion as object presence. Deep learning techniques improve segmentation but still lose fine object details. The proposed method achieves better delineation and consistently detects objects despite occlusion or environmental disturbances.

Figure 4 presents qualitative results on the Underwater Change Detection dataset, which includes five challenges: caustics, fish swarm, marine snow, small aquaculture, and two fishes. The proposed method performs consistently well across all cases. Despite real-world complexities such as low lighting and color distortion, the model successfully detects relevant objects with high fidelity.

## 6 Quantitative Analysis

We conducted a quantitative comparison between our model and a set of twelve benchmark techniques to assess its performance. As shown in Table 1, conventional algorithms such as GMM [3], KDE [4], and ML-BGS [31] perform poorly under complex or dynamic conditions. Deep learning approaches like FgSegNet [32] and U-UNet [8] provide improvements but still fall short in maintaining object boundaries.

Graph-based models such as GraphSage [12] demonstrate stronger performance, although limitations are observed in crowded scenes. The proposed hypergraph based model achieves the highest F-measure scores across all scenarios. This improvement is attributed to the ability of hypergraphs to model higher-order relationships, thereby better representing the latent structure of complex scenes.

Table 2 extends this analysis to the Underwater Change Detection dataset. The proposed method consistently outperforms thirteen other approaches across all five challenges, achieving the highest average F-measure. These results further confirm the robustness of our approach.

## 7 Ablation Study

Two ablation studies were conducted to assess the effect of architectural choices. Table 3 summarizes and compares the use of convolutional versus attention-based kernels. Results show that while attention mechanisms can improve certain metrics, they may also introduce noise when applied to the incidence matrix. As such, a standard convolution kernel offers a more stable and efficient choice for this task.

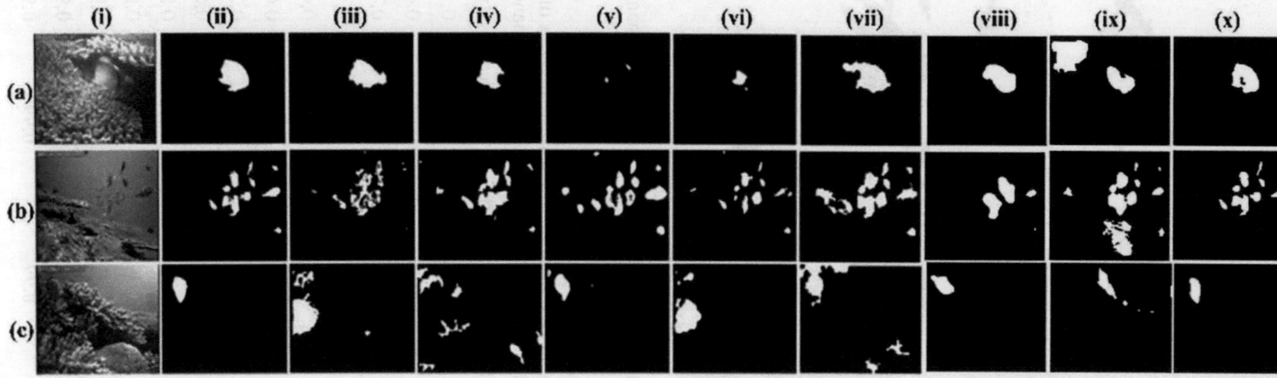


**Fig. 3** Visual analysis of proposed hypergraph refactoring with deep learning based methods on Fish4Knowledge dataset. Here, (i) input image, (ii) ground-truth, (iii) Texture-BGS, (iv) ML-BGS, (v) MultiCueBGS, (vi) SuBSENSEBGS, (vii) SILTP, (viii) FgSegNet, (ix) U-UNet, (x) Proposed method. Three challenges of the dataset are considered: **a** complex background, **b** crowded scenes, **c** dynamic background

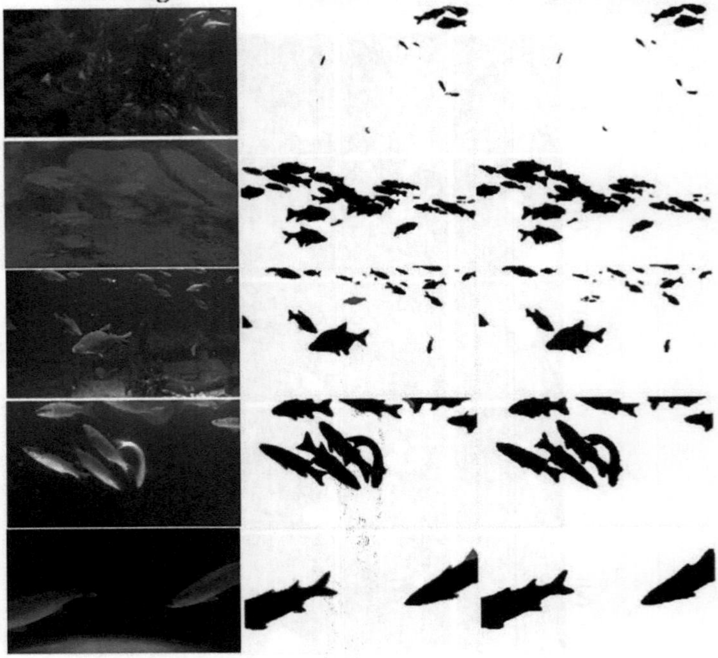


**Fig. 4** Visual analysis of proposed hypergraph refactoring on underwater change detection dataset, evaluated on five challenges

**Table 1** Quantitative analysis of proposed hypergraph refactoring in terms of average F-Measure on four challenges of Fish4Knowledge dataset. Bold indicates best performance

| Method | ComplexBkg | DynamicBkg | Crowded | Hybrid | Overall |
|---|---|---|---|---|---|
| GMM [3] | 0.12 | 0.06 | 0.26 | 0.15 | 0.15 |
| ML-BGS [31] | 0.58 | 0.32 | 0.74 | 0.46 | 0.52 |
| MC-BGS [28] | 0.48 | 0.33 | 0.68 | 0.72 | 0.55 |
| KDE [4] | 0.13 | 0.07 | 0.20 | 0.16 | 0.14 |
| SuBSENSEBGS [33] | 0.21 | 0.81 | 0.67 | 0.42 | 0.53 |
| FgSegNet [32] | 0.64 | 0.39 | 0.68 | 0.60 | 0.58 |
| U-Unet [8] | 0.31 | 0.27 | 0.23 | 0.22 | 0.26 |
| MFI [6] | 0.83 | 0.64 | 0.69 | 0.64 | 0.70 |
| CSLTP [5] | 0.80 | 0.76 | – | 0.83 | 0.79 |
| MSGAN [10] | 0.92 | 0.79 | 0.74 | 0.89 | 0.83 |
| HMLS [34] | 0.84 | 0.90 | 0.84 | 0.91 | 0.87 |
| GraphSage [12] | **0.99** | **0.99** | **0.98** | **0.98** | **0.99** |
| Proposed | **0.99** | **0.99** | **0.99** | **0.98** | **0.99** |

**Table 2** Quantitative analysis in terms of F-measure for thirteen SOTA methods across five challenges of underwater change detection. Bold indicates the best performance

| Models | Caustics | Fish Swarm | Marine snow | Small aquaculture | Two fishes | Avg. |
|---|---|---|---|---|---|---|
| ABMM [35] | 0.06 | 0.65 | 0.43 | 0.67 | 0.76 | 0.51 |
| AGMM [36] | 0.30 | 0.82 | 0.74 | 0.74 | 0.79 | 0.68 |
| GSMM [37] | 0.57 | 0.84 | 0.77 | 0.55 | 0.79 | 0.70 |
| ADE [38] | 0.59 | 0.82 | 0.88 | 0.75 | 0.71 | 0.75 |
| GWFT [39] | 0.85 | 0.91 | 0.93 | 0.67 | 0.82 | 0.84 |
| SSSR [40] | 0.79 | 0.80 | 0.83 | 0.85 | 0.91 | 0.83 |
| DeColor [41] | 0.69 | 0.72 | 0.73 | 0.81 | 0.84 | 0.75 |
| SRPCA [42] | 0.72 | 0.73 | 0.75 | 0.80 | 0.82 | 0.76 |
| MSCL [43] | 0.75 | 0.74 | 0.80 | 0.83 | 0.89 | 0.80 |
| GFL [44] | 0.77 | 0.75 | 0.76 | 0.77 | 0.84 | 0.75 |
| 2PRPCA [45] | 0.71 | 0.76 | 0.72 | 0.74 | 0.82 | 0.70 |
| OMoGMF + TV [46] | 0.66 | 0.70 | 0.70 | 0.67 | 0.79 | 0.70 |
| CS-RPCA [47] | 0.81 | 0.83 | 0.85 | 0.88 | 0.95 | 0.86 |
| Proposed | **0.99** | **0.99** | **0.99** | **0.98** | **0.99** | **0.99** |

**Table 3** Ablation study with and without attention in hypergraphs. Metrics: Acc. = Accuracy, Rec. = Recall, Prec. = Precision, F1 = F-measure

| | Without attention | | | | With attention | | | |
|---|---|---|---|---|---|---|---|---|
| | Acc. | Rec. | Prec. | F1 | Acc. | Rec. | Prec. | F1 |
| DynamicBkg | 98.25 | 98.26 | 99.99 | 99.12 | 98.35 | 98.39 | 99.96 | 99.17 |
| Complexbkg | 98.41 | 98.51 | 99.90 | 99.20 | 98.68 | 98.80 | 99.87 | 99.33 |
| Crowded | 99.21 | 99.39 | 99.82 | 99.61 | 99.22 | 99.52 | 99.70 | 99.61 |
| Hybrid | 96.62 | 96.72 | 99.88 | 98.28 | 94.10 | 96.77 | 97.16 | 96.96 |
| Standard | 97.91 | 97.99 | 99.91 | 98.94 | 97.99 | 98.37 | 99.60 | 98.98 |
| Lumichange | 99.76 | 99.85 | 99.92 | 99.88 | 99.74 | 99.83 | 99.91 | 99.87 |
| Overall | 98.36 | 98.46 | 99.90 | 99.17 | 98.01 | 98.61 | 99.37 | 98.99 |

## 8 Discussions and Future Works

This work introduces a graph-refactored strategy in latent space that improves information retention in CNN-based architectures. Earlier approaches primarily relied on pairwise connections between nodes, which often failed to model complex relationships in dynamic underwater environments. The proposed method incorporates hyperedges to model higher-order dependencies, leading to improved structural preservation and robustness.

Despite these gains, the model has limitations in generalizing to completely unseen environments. Performance degradation may occur in scenarios that differ significantly from the training distribution. Future work will focus on developing a semi-supervised learning framework to improve adaptability by leveraging both labeled and unlabeled data. This extension is expected to enhance the model's generalization capabilities for real-world underwater object detection tasks.

## 9 Conclusions and Future Works

This chapter introduced a novel end-to-end graph-refactored encoder-decoder architecture tailored for underwater object detection, designed with the broader objective of integration into Digital Twin (DT) systems for marine environments. By embedding hypergraph-based reasoning within the latent space, the model effectively captures complex spatial and contextual relationships that are critical for accurately representing the dynamic underwater world. An unsupervised graph initialization strategy was proposed to address the absence of labeled node information, enabling adaptive graph construction suitable for real-world DT scenarios where labeled data may be limited or unavailable. Additionally, we examined message passing techniques using both convolutional and attention-based kernels, identifying convolution as a more reliable mechanism for preserving structural information in noisy underwater data. Extensive evaluation against fourteen state-of-the-art methods demonstrated the robustness and accuracy of the proposed approach. The ability to preserve fine-grained object details, even in challenging conditions such as occlusion, dynamic backgrounds, and visual distortions, positions this method as a strong candidate for deployment within digital replicas of marine environments. Looking ahead, this work opens avenues for enhancing DT frameworks with more intelligent visual processing modules. Future research will focus on multi-modal integration, semi-supervised learning, and real-time deployment to further align the system with the operational needs of Digital Twins in marine monitoring, ecological assessment, and autonomous underwater navigation [48–51].

## References


1. Subudhi BN et al (2022) Kernel-induced possibilistic fuzzy associate background subtraction for video scene. IEEE Trans Comput Soc Syst 10(3):1314–1325
2. Rout DK et al (2024) Underwater visual surveillance: a comprehensive survey. Ocean Eng 309:118367
3. Stauffer C, Grimson WEL (2000) Learning patterns of activity using real-time tracking. IEEE Trans Pattern Anal Mach Intell 22(8):747–757

4. Elgammal A et al (2002) Background and foreground modeling using nonparametric kernel density estimation for visual surveillance. Proc IEEE 90(7):1151–1163
5. Vasamsetti S et al (2019) 3D local spatio-temporal ternary patterns for moving object detection in complex scenes. Cogn Comput 11:18–30
6. Vasamsetti S et al (2018) Automatic underwater moving object detection using multi-feature integration framework in complex backgrounds. IET Comput Vis 12(6):770–778
7. Rout DK et al (2018) Spatio-contextual Gaussian mixture model for local change detection in underwater video. Expert Syst Appl 97:117–136
8. Bajpaietal V (2021) Underwater U-Net: Deep learning with U-Net for visual underwater moving object detection. In: Proceedings of the OCEANS: San Diego–Porto. IEEE. pp 1–4
9. Li X et al (2015) Fast accurate fish detection and recognition of underwater images with fast R-CNN. In: Proceedings of the OCEANS, Washington. IEEE. pp 1–5
10. Patil PW et al (2019) Motion saliency based generative adversarial network for underwater moving object segmentation. In: Proceedings of the IEEE international conference on image processing (ICIP), pp 1565–1569
11. Giraldo JH, Javed S, Bouwmans T (2020) Graph moving object segmentation. IEEE Trans Pattern Anal Mach Intell 44(5):2485–2503
12. Kapoor M et al (2023) Underwater moving object detection using an endto-end encoder-decoder architecture and GraphSage with aggregator and refactoring. In: Proceedings of the IEEE/CVF conference on computer vision and pattern recognition, pp 5635–5644
13. Wiatowski T, Bolcskei H (2017) A mathematical theory of deep¨ convolutional neural networks for feature extraction. IEEE Trans Inform Theory 64(3):1845–1866
14. Rout DK et al (2017) A novel five-frame difference scheme for local change detection in underwater video. In: Proceedings of the IEEE fourth international conference on image information processing (ICIIP), pp 1–6
15. Prabowo MR et al (2017) A moving objects detection in underwater video using subtraction of the background model. In: Proceedings of the international conference on electrical engineering, computer science and informatics (EECSI), pp 1–4
16. Pagire V, Phadke A (2021) Underwater moving object detection using GMG. In: Proceedings of the international conference on soft computing and pattern recognition. Springer, pp 233–244
17. Salman A et al (2019) Real-time fish detection in complex backgrounds using probabilistic background modelling. Eco Inform 51:44–51
18. Fu C et al (2023) Learning heavily-degraded prior for underwater object detection. IEEE Trans Circ Syst Video Technol 33(11):6887–6896
19. Song P et al (2023) Boosting R-CNN: reweighting R-CNN samples by RPN's error for underwater object detection. Neurocomputing 530:150–164
20. Tseng S-L, Lin H-Y (2020) Fish detection using convolutional neural networks with limited training data. In: Proceedings of Asian conference on pattern recognition. Springer, pp 735–748
21. Sun Y et al (2022) Underwater object detection with Swin transformer. In: Proceedings of the international conference on data intelligence and security (ICDIS), pp 422–427
22. Liu J et al (2022) Two-stage underwater object detection network using Swin transformer. IEEE Access 10:117235–117247
23. Xu H et al (2019) Spatial-aware graph relation network for large-scale object detection. In: Proceedings of the IEEE/CVF conference on computer vision and pattern recognition, pp 9298–9307
24. Chen S, Li Z, Tang Z (2020) Relation R-CNN: a graph based relation-aware network for object detection. IEEE Signal Process Lett 27:1680–1684
25. Liang Z et al (2021) Visual-semantic graph attention networks for human object interaction detection. In: Proceedings of IEEE international conference on robotics and biomimetics (ROBIO). IEEE, pp 1441–1447
26. He Q et al (2022) SVGA-Net: sparse voxel-graph attention network for 3d object detection from point clouds. In: Proceedings of the AAAI conference on artificial intelligence, vol 36, issue 1, pp 870–878

27. Bai S, Zhang F, Torr PHS (2021) Hypergraph convolution and hypergraph attention. Pattern Recogn 110:107637
28. Noh SJ, Jeon M (2013) A new framework for background subtraction using multiple cues. In: Proceedings of the Asian conference on computer vision. Springer, pp 493–506
29. St-Charles P-L, Bilodeau G-A, Bergevin R (2017) Flexible background subtraction with self-balanced local sensitivity. In: Proceedings of the IEEE conference on computer vision and pattern recognition workshops, pp 408–413
30. Heikkila M, Pietikainen M (2006) A texture-based method for modeling the background and detecting moving objects. IEEE Trans Pattern Anal Mach Intell 28(4):657–662
31. Yao J, Odobez J-M (2007) Multi-layer background subtraction based on color and texture. In: Proceedings of the IEEE/CVF conference on computer vision and pattern recognition, pp 1–8
32. Lim LA, Keles HY (2018) Foreground segmentation using convolutional neural networks for multiscale feature encoding. Pattern Recogn Lett 112:256–262
33. St-Charles P-L, Bilodeau G-A, Bergevin R (201) SuBSENSE: a universal change detection method with local adaptive sensitivity. IEEE Trans Image Process 24(1):359– 373
34. Salman A et al (2020) Automatic fish detection in underwater videos by a deep neural network-based hybrid motion learning system. ICES J Marine Sci 77(4):1295–1307
35. Kaewtrakulpong P, Bowden R (2002) An improved adaptive background mixture model for real-time tracking with shadow detection. In: Video-based surveillance systems: computer vision and distributed processing, pp 135–144
36. Zivkovic Z (2004) Improved adaptive Gaussian mixture model for background subtraction. In: Proceedings of the IEEE international conference on pattern recognition, vol 2, pp 28–31
37. Radolko M, Gutzeit E (2015) Video segmentation via a Gaussian switch background model and higher order Markov random fields. In: International conference on computer vision theory and applications, pp 537–544
38. Zivkovic Z, Van Der Heijden F (2006) Efficient adaptive density estimation per image pixel for the task of background subtraction. Pattern Recogn Lett 27(7):773–780
39. Radolko M, Farhadifard F, von Lukas U (2017) Change detection in crowded underwater scenes-via an extended Gaussian switch model combined with a flux tensor pre-segmentation. In: Proceedings of the international conference on computer vision theory and applications, vol 5, pp 405–415
40. Javed S et al (2018) Moving object detection in complex scene using spatiotemporal structured-sparse RPCA. IEEE Trans Image Process 28(2):1007–1022
41. Zhou X, Yang C, Yu W (2012) Moving object detection by detecting contiguous outliers in the low-rank representation. IEEE Trans Pattern Anal Mach Intell 35(3):597–610
42. Javed S et al (2016) Spatiotemporal low-rank modeling for complex scene background initialization. IEEE Trans Circ Syst Video Technol 28(6):1315–1329
43. Javed S et al (2017) Background–foreground modeling based on spatiotemporal sparse subspace clustering. IEEE Trans Image Process 26(12):5840–5854
44. Xin B et al (2015) Background subtraction via generalized fused lasso foreground modeling. In: Proceedings of the IEEE conference on computer vision and pattern recognition, pp 4676–4684
45. Gao Z, Cheong L-F, Wang Y-X (2014) Block-sparse RPCA for salient motion detection. IEEE Trans Pattern Anal Mach Intell 36(10):1975–1987
46. Yong H et al (2017) Robust online matrix factorization for dynamic background subtraction. IEEE Trans Pattern Anal Mach Intell 40(7):1726–1740
47. Javed S et al (2020) CS-RPCA: clustered sparse RPCA for moving object detection. In: Proceedings of the IEEE international conference on image processing. IEEE, pp 3209–3213
48. Sharma L, Carpenter M (eds) (2022) Computer vision and Internet of Things: technologies and applications, 1st edn. Chapman and Hall/CRC. https://doi.org/10.1201/9781003244165
49. Sharma L, Garg PK (eds) (2021) Artificial intelligence: technologies, applications, and challenges, 1st edn. Chapman and Hall/CRC. https://doi.org/10.1201/9781003140351
50. Sharma L (ed) (2020) Towards smart world: homes to cities using Internet of Things, 1st edn. Chapman and Hall/CRC. https://doi.org/10.1201/9781003056751

51. Sharma L, Garg PK (eds) (2019) From visual surveillance to Internet of Things: technology and applications, 1st edn. Chapman and Hall/CRC. https://doi.org/10.1201/9780429297922

# Chapter 11
# A Smart Framework for Early Crop Disease Detection Using GIS Digital Twin Technology

**Sudhriti Sengupta, Lavanya Sharma, and Laxmi Ahuja**

**Abstract** Early identification of crop diseases is critical to ensure food security by reducing yield losses and promoting sustainable agricultural practices. Advanced image processing and artificial intelligence methods are employed for analysing visual symptoms and detecting anomalies indicative of disease onset. Digital twin environment supports informed decision-making and timely responses by simulating disease progression and assessing various intervention strategies. This chapter discusses the methodologies used in plant health monitoring. It also elaborates the concept of digital twins, the technologies that enable them, and their key components. This chapter presents a smart framework that utilizes digital twin technology to detect preliminary stage of plant diseases. The proposed system creates a real-time virtual replica of the agricultural environment by integrating data streams originating from sensors, drones, and interconnected IoT systems.

**Keywords** Plant disease detection · Food safety · Digital twins · Image processing technology

## 1 Introduction

Detecting plant diseases is essential for sustaining agricultural productivity. Through the implementation of preventative measures, farmers can restrict the spread of illnesses and lower the risk of crop losses when diseases are detected early. Detecting plant diseases safely and effectively is vital for maintaining healthy crops, ensuring

---

S. Sengupta (✉) · L. Ahuja
Amity Institute Of Information Technology, Amity University, Noida, Up, India
e-mail: sudhriti919@gmail.com

L. Ahuja
e-mail: lahuja@amity.edu

L. Sharma
Department Of Information Technology, New Delhi Institute Of Management, Delhi, India

 
L. Sharma et al. (eds.), *Digital Twin Technology in Geospatial Science: Trends, Applications and Challenges*, Data-Intensive Research,
https://doi.org/10.1007/978-3-032-20311-3_11

food security and supporting livelihood, particularly in areas where agriculture is the main source of income [1]. It enables the most efficient use of resources by applying pesticides and fertilizers specifically where they need to be applied. This lowers the production cost for the farmers and prevents surplus usage of chemicals. Further, timely and correct detection of plant disease aids the agriculturists in controlling crops yield by making the growth more adaptable against disease outbreak. This promises to promote reliable and dependable global food supply by ensuring healthy crop production and successful supply chain management of products to the consumers. Efficient and effective plant disease detection supports the farmers in guaranteeing proper harvest leading to economic stability, food security and long-term sustainability [2].

Unhealthy plants display symptoms like discoloured patches and spots on leaves, stems, fruits or flowers. These signs make it easier for timely diagnosis of plant diseases and mitigate them. For example, yellowing or wilting of leaves typically suggests fungal disease, whereas a viral infection or nutrition deficiency can be identified by dark spots. Traditionally, detection of plant disease strongly relied on assessment by agricultural experts or experience of farmers. This method was slow, labour intensive and sometimes subjective leading to mistakes. Wrong assessments of plant disease led to improper application of pesticides, leading to decrease in quality of crop. To overcome these challenges, utilization of image processing techniques for identification of plant disease has become an active area of research.

Apart from ensuring food security, effective diagnosis and mitigation of crop diseases are also important for protecting native plant species and keeping ecosystems stable researchers in agriculture can develop innovative varieties of plants that are more resistant to disease and enhance farming methods by understanding disease patterns and trends [3, 4]. The benefits of detecting crop diseases timely and accurately can be described as:

Early detection prevents large-scale crop losses, helping maintain stable food supplies and thus ensuring food security.

- Timely identification and treatment of diseases boost yield and crop quality which improves agricultural productivity.
- Prevents revenue loss for farmers and reduces costs related to pesticide overuse and crop failure.
- Enables targeted disease management, reducing unnecessary chemical use and promoting eco-friendly practices which improves public health.
- Helps preserve native plant species and prevent the spread of diseases to wild plant populations and protect plant based eco system.
- Facilitates research and breeding programs in developing disease-resistant plant varieties and improving farming practices.
- Early warning for emerging threats enables rapid response to new or invasive plant pathogens, limiting their spread.

Researchers have been working on developing computer-aided diagnostics for plant disease by application of both traditional image processing and advanced artificial intelligence methods to enable plant disease identification and management.

Computer-aided diagnostics system promises to provide more effective identification as it can identify subtle visual signs (spots, discolouration, texture changes) before they are visible to the human eye. Also automatized system can eliminate human errors by applying consistent evaluation criteria across all samples. Automatization provides faster diagnosis, strong accuracy and dependability of disease detection. Integration of satellite imaginary, drone, IoT sensors and smart device supports real-time and larger collection of data as compared to manual intervention. Also researchers can take immense benefits from the stored records for long-term analysis and tracking disease progression over time.

Besides the conventional image processing and artificial intelligence-based techniques, various state of the art concepts are also advancing in created of automatized plant disease detection system. One of these emerging technologies is the concept of digital twins. This technology is used to create virtual replica of agricultural systems that mimic real-world circumstances. Digital twin models integrate data from different sources like sensors, IoT devices, drone, satellite images to record, monitor and predict disease outbreak. Agriculturists and farmers can experiment with different management techniques and predict risks by examining the simulated surroundings without endangering real crops.

## 2 Plant Disease

Plant disease refers to any abnormal physiological or morphological condition in induced by microbial pathogens such as fungi, bacteria, and viruses, nematodes, or by adverse climatic variables. These diseases disrupt normal plant functions such as photosynthesis, nutrient uptake, or reproduction causing symptoms like wilting, chlorosis, necrosis, stunted growth, and reduced yield or quality [23–25]. Plant diseases are typically categorized on basis of factors causing them (biotic/abiotic), parts of infected plants and symptoms which they show. Biotic diseases are more infectious and caused by living organisms whereas abiotic diseases are caused by non-living factors such as deficiency of nutrients, pollutions and environmental effects. Timely detection and effective management of plant disease are essential to maintain crop production, protect food security and reduce financial loss. Advanced technologies, viz, remote sensing, image processing and digital twin technology, are being used to enhance disease diagnosis, monitoring and control methods in modern agriculture.

- **Leaf Spots and Lesions**: Browning, blackening, or necrotic areas on leaves are typical of fungal or bacterial infections. These often feature darker borders or centres with lighter halos. Examples include bacterial spots with yellow halos and fungal spots with spore-bearing centres.
- **Powdery Mildew**: White or greyish powdery coating across leaf surfaces, indicative of fungal pathogens in *Erysiphales*. Most common on stems and lower leaves, where mycelia can be rubbed off.

- **Rust Diseases**: Characterized by orange, yellow, or reddish pustules (uredinia) disrupting photosynthesis. The powdery clusters often break open, weakening the leaf tissue.
- **Viral Symptoms (Mosaic & Distortion)**: Mottling (patchy yellow and green), chlorosis, ringspots, and leaf distortion are hallmarks of viral infections (e.g., CMV). Plant leaves may appear deformed with mosaic patterns.

## 3 Methodologies Used in Plant Disease Detection

The initial stage of traditional image processing approach applied for plant disease detection starts with image input, where images of plants modalities are acquired using tools like cameras, drones, or satellites, emphasizing the importance of quality and resolution. This is followed by image pre-processing, which improves the images quality by applying methods like noise removal, contrast enhancement, resizing, filtering and normalization. The next stage is feature extraction where specific characteristics like colour, texture, and patterns are identified, as area of interest to detect plant condition or occurrence of pests and disease. Finally, the classification step utilizes the area of interest to categorize the images into healthy or diseased plants, often methods of artificial intelligence. Overall, this systematic approach leverages technology for improving and effectiveness of plant disease detection in agricultural practices [5, 6]. The flowchart as given in Fig. 1 depicts this approach of image processing used in diagnosis of plant disease.

### *3.1 Image Input*

Image input is one of the important stages in development of the automated detection method of plant diseases. Detailed images of plant parts, like leaves, stems, flowers, etc., are captured using digital cameras, smartphones, or specialized imaging sensors. These images act as main data input for next phases in image processing techniques. Accurate image input is essential because the quality of the images directly affects the effectiveness of disease identification. By collecting consistent and detailed visual data, image input led to early detection of symptoms such as spots, discolouration, or lesions which supports timely and precise disease diagnosis in agricultural systems. GIS (geographic information system) improves image input for plant disease detection by connecting spatial data with visual symptoms, enabling location-based

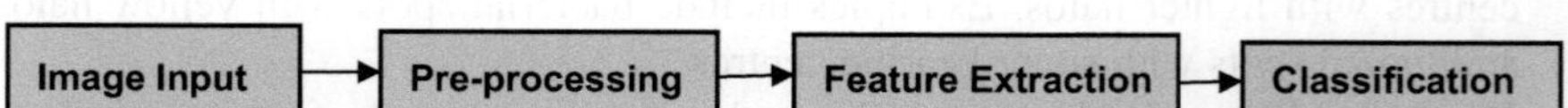


**Fig. 1** Key steps in image processing for detecting plant diseases

analysis of disease spread. By integrating geotagged images, GIS supports precise monitoring and mapping of plant diseases across large agricultural areas.

## 3.2 Image Pre-processing

The quality and perceivability of images is improved by using image pre-processing steps. Some of the common pre-processing techniques used in plant disease detection are filtering, resizing, contrast enhancement, noise removal, etc. Filtering techniques are used to remove artifices and unnecessary details, like dirt, dust, water spot, etc. Resizing images is used for standardizing input data, ensuring consistent aspect ratio and preserving important details to get consistency in next stages of image analysis. Contrast enhancement is used to increase the visibility of area of interest by adjusting the lighting factors. These pre-processing stages are utmost important as accurate and reliable data can be obtained after enhancing image quality using different traditional of artificial intelligence-based methods. The GrabCut algorithm is widely used for background removal because of its efficiency and precision in object segmentation. This algorithm requires manual selection of the region of interest (ROI) which is challenging in case of intricate patterns. To overcome this challenge, researchers have analysed various hybrid methods that combine GrabCut with other segmentation techniques like DeepLabV3 + , Canny edge detection and the watershed algorithm [7].

To improve image quality, researchers have applied many techniques. Some of the most popular ones are interpolation techniques used for increasing image resolution, filtering techniques for noise removal, histogram equalization methods for enhancing contrast, edge detection operators for border detection, etc. These techniques [8] successfully address challenges such as low image quality and inconsistent lighting for improving image quality to enable better disease diagnosis.

## 3.3 Feature Extraction

Feature extraction represents a critical step in plant disease detection, as it involves identifying and isolating key characteristics from images that distinguish healthy plants from diseased ones. These features often includes colour, texture, shape, and edge patterns of leaves, stem, fruits or other modalities. Effective feature extraction improves diagnostic precision and enhances the performance of automated disease recognition systems. Some of the statistical methods used for extraction of textures are local binary patterns (LBP), grey level co-occurrence matrix (GLCM), colour co-occurrence matrix (CCM), and spatial grey level dependence matrix (SGLDM). Additionally, model-based approaches such as auto-regressive (AR) and Markov random field (MRF) models can also be used to extract these features. Several types of wavelet transforms including discrete wavelet transforms (DWT), Haar wavelets, and Daubechies wavelets [9] are also popular in feature extraction in plant disease automation.

### *3.4 Classification*

Image classification is a key component in automatization processes categorizing healthy plants from diseased one. It is generally applied after the pre-processing and feature extraction staged. Classification can be also used subsequently for cat goring disease types and subtypes. A diverse and large set of images that represent the various categories is required to train a robust model for classification in various types. An appropriate model such as convolutional neural networks (CNNs) and pre-trained models (e.g., VGG16, ResNet, etc.) is chosen and trained with the diverse and large training set of images. The performance of this model is assessed using metrics such as accuracy, precision, recall, and F1-score. Generally, confusion matrices are used to understand where the model is making errors. Yang et al. [10] proposed a deep learning-based approach for classifying plant diseases using an enhanced channel attention mechanism. In 2023, Saad and Salman [11] proposed a one-shot learning (OSL) technique for classifying plant diseases using Siamese neural networks (SNN) with limited datasets. In 2024, Shaqik et al. [12] developed a methodology integrating early fusion with lead voting ensemble techniques, employing nine pre-trained CNNs optimized for deep feature extraction.

## 4 Digital Twins

Digital twin is a virtual representation of a physical system, process, or thing that mimics its behaviour, performance, and attributes in real time. This concept is often applied to enhance decision-making processes and streamline operations in a variety of sectors, including manufacturing, healthcare, urban planning, and more. In manufacturing sector, digital twins can be used for production lines monitoring, optimizing supply chains, and improving equipment maintenance [13]. Creating digital twins of patients or medical devices to personalize treatment plans and enhance outcomes plays a promising role in advancement of patient care and treatment. In smart cities managing urban infrastructure through simulations of traffic flow, energy consumption, and resource distribution can be improved using these techniques [26–29].

Figure 2 depicts the main components of digital twins

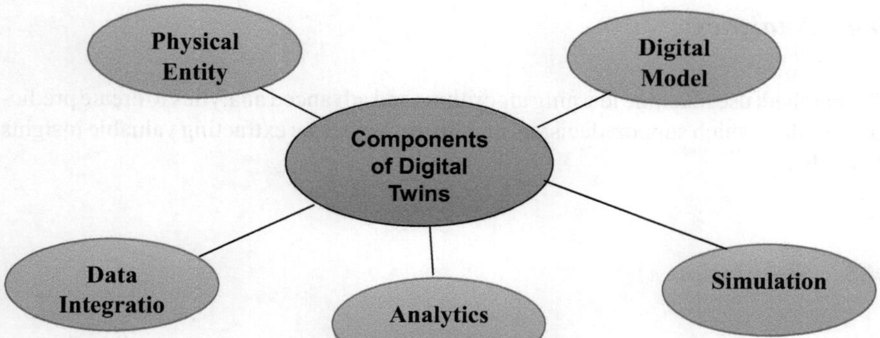


**Fig. 2** Digital twins components

### *4.1 Physical Entity*

Physical entity represents a real-world object, system, or process that it possibly replicates, such as a machine, a building, or even an entire production facility.

### *4.2 Digital Model*

A detailed virtual representation that represents the structure, behaviour, and attributes of a physical entity, enabling simulation, analysis, and monitoring in a digital environment is the digital model. The main components of digital model are sensor data, historical data and 3D models.

### *4.3 Data Integration*

Data integration is the process of collecting, integrating, and synchronizing data from physical systems with their digital counterparts. With the help of sensors and internet of things technologies, it enables constant monitoring and analysis. Effective data integration guarantees real-time accuracy and promotes data-driven decision-making.

### 4.4 Analytics

Researchers use machine learning algorithms and advanced analytics to create prediction models which support decision-making processes by extracting valuable insights from data.

### 4.5 Simulation

Simulation is the process of emulating numerous scenarios to assess possible changes and predict how the physical system will react to particular situations. By real-time monitoring and performance analysis, organizations can identify inefficiencies and optimize operations accordingly. Digital twins can predict when equipment is likely to fail, allowing for predictive maintenance and reduced downtime. Ensuring data security and seamless integration is one of the chief concern in usage of digital twins in various domains.

In 2024, Dia et al. [14] proposed a digital twin system by integrating data and model fusion techniques for pest management. At first, a comprehensive digital twin framework is proposed to monitor and control insect pests throughout the crop growth cycle. Subsequently, a predictive digital twin model is developed using a random forest algorithm optimized through a genetic algorithm to estimate pest populations. Digital twins in plant disease prediction and analysis serve as dynamic virtual models of crop systems that integrate real-time sensor data, environmental conditions, and historical records to simulate plant health, detect early signs of disease, and support accurate diagnosis and targeted intervention strategies.

## 5 Technologies in Digital Twins

Digital twin enables creation of dynamic, data-driven virtual representations of physical assets by merging different technologies. When combined, these technologies make it possible to capture data in real time and model, simulate, and analyse real-world environments [15, 21–23]. These facilitate better monitoring, well-informed decision-making, and forecasted assessment of crop health and field conditions in agriculture. Some of the common techniques are mentioned below:

- Internet of things (IoT) captures real-time data from physical environments. The main function of IoT is to monitor soil moisture, temperature, humidity, and other plant health indicators using sensors, actuators, etc.
- Remote sensing and drones assists in acquiring high-resolution spatial imagery which enables detection of early visual symptoms (e.g., leaf discolouration, lesions) from above.

- Geographic information systems (GIS) map the spatial distribution of infected crops and monitor the progression of diseases across the agricultural landscape.
- Artificial intelligence and machine learning (AI/ML) process patterns, categorize conditions, and generate predictive insights, enabling the identification of disease symptoms from images and the forecasting of outbreaks based on historical data.
- Big data analytics processes and extracts insights from large, heterogeneous datasets.
- Cloud computing offers scalable data storage, processing, and access anywhere and anytime.
- Simulation and modelling tools can simulate biological and environmental interactions, models crop growth, pest dynamics, and responses to interventions such as irrigation or fertilization.
- Cybersecurity technologies protects data integrity and privacy by ensures secure transmission and storage of sensor and farm management data.

These technologies operate together in the digital twin framework to monitor, identify, simulate, and predict plant diseases, which helps timely interventions to reduce crop losses.

## 6 Proposed Framework for Plant Disease Detection Using Digital Twin

The proposed digital twin framework for agricultural landscapes integrates real-time data and spatial information to create a dynamic, virtual representation of farming environments. Central to this framework is the use of geographic information system (GIS) maps, which provide spatially accurate and layered visualization of agricultural fields, including topography, soil types, crop distributions, and infrastructure. Digital twins integrates data collected by remote sensing, IoT devices, drones and sensors to replicate virtual copy of the physical field conditions. The virtual model helps in precision farming by letting agriculturists to visually understand the components of the landscape and predict future changes [16–20]. Through continuous updates and analysis, it improves awareness of the field conditions enabling better resource management leading to sustainable agricultural practices (Fig. 3).

This framework applies the GIS maps to provide a detailed multi-tiered and geospatial visualization of agricultural area. These maps incorporate crucial landscape features, including terrain topography, soil composition, crop types and distributions, water resources, irrigation networks, and farm infrastructure. Sensors form a crucial part of main data input. They provide vital information about the environmental conditions, like temperature, humidity level, etc. To maintain an up-to-date and realistic simulation of field conditions, the proposed framework continuously synchronizes heterogeneous data streams from these various sources. These inputs enable the system to reflect current environmental conditions (e.g., soil moisture,

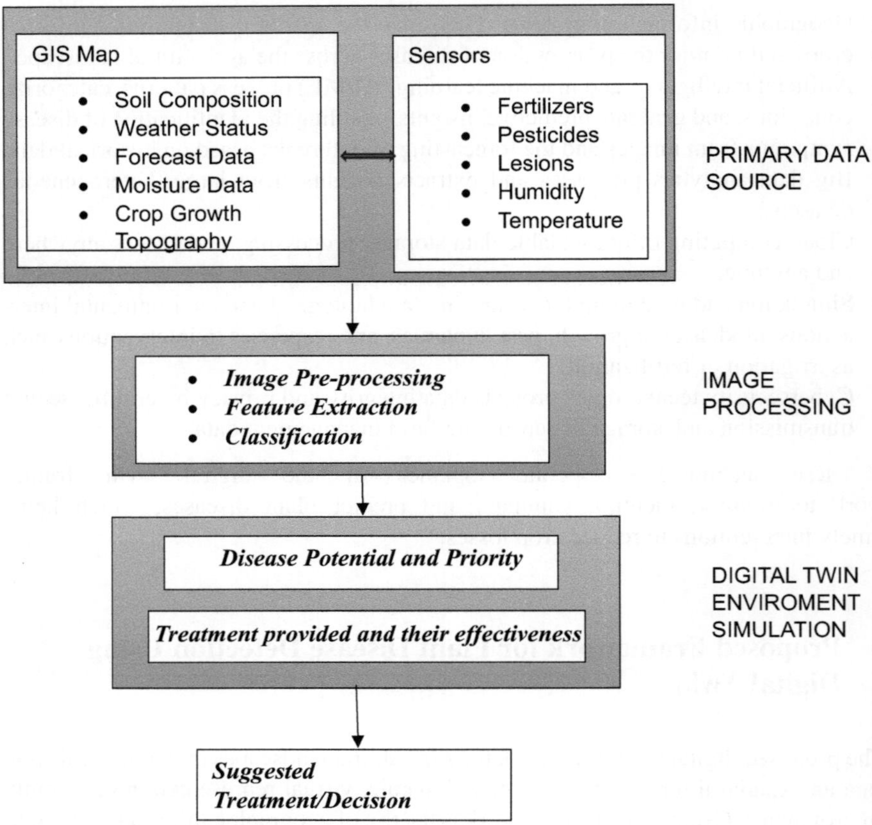


**Fig. 3** A proposed framework for plant disease detection

temperature, humidity, pest activity) and support predictive modelling for future scenarios like disease outbreaks, yield estimates, or climate impacts.

After data collection from various heterogeneous sources, the conventional steps in image processing like image pre-processing, feature extraction and classification is applied to determine if any area is affected by disease or not. A step beyond these traditional techniques can be achieved by simulation offered using digital twins environments. It is possible to map the physical counterpart of the diseased plants to digital models and use advanced analytics and simulation techniques to observe the capacity of disease to grow in future. The digital twin techniques through various analytical predictive models can determine the most efficient and cost-effective treatment in future.

This geospatially enabled framework enhances agricultural decision-making by enabling stakeholders—farmers, agronomists, and policymakers—to virtually explore and interact with the digital environment. To make this geospatial framework more efficient in decision-making, the images require application of pre-processing techniques. The right image pre-processing technique makes data clearer, analysis more accurate and decisions more reliable.

- ***Data Quality Improvement***: Pre-processing techniques such as noise reduction, contrast enhancement, and filtering improve the quality of images. High-quality images lead to better understanding and more accurate results.
- ***Feature Extraction***: Pre-processing assists in detecting and highlighting main features like types of soil, crops and water sources. These findings are very essential to the stakeholders to study agricultural field more accurately.
- ***Normalization and Standardization***: Consistent images across different sensors and conditions ensure that data can be compared. This allows the creation of different test cases for a variety of scenarios and conditions.
- ***Enhanced Visualization***: Consistent images across different sensors and conditions ensures that the data can be compared across different test cases for a variety of scenarios and conditions.

Methods such as Gaussian and median filtering are utilized to remove noise and improve clarity of the images. Contrast stretching, histogram equalization and sharpening improve visual features for better analysis and perceivability. Operations such as rotating and cropping align and resize all images so that they can be consistent. There are many conventional image pre-processing techniques which can be used for enhancing image qualities. Amongst all the image pre-processing techniques, contrast enhancement techniques are one of the most popular methods used. Contrast enhancement improves the visibility of features in images, making details more perceptible. It improves overall quality and clarity, which facilitates better interpretation and analysis. Contrast enhancement techniques like intensity adjustment, histogram equalization, and adaptive histogram equalization are widely used in image pre-processing. This section provides an experimental comparison of these three contrast enhancement techniques to determine the most efficient one. This aims to leverage the images obtained from various sources as per proposed framework for better perception (Table 1).

Entropy of image is used to determine the quantitative parameter of image quality. These output images are compared by using the entropy values of image as per the different techniques.

A graphical representation of Table 2 is shown below to depict that entropy of image obtained by adaptive histogram equalization is significantly higher than that of entropy values of image obtained by other techniques (Fig. 4).

In a digital twin framework for plant disease detection, image contrast plays a crucial role in enhancing the visibility of features that indicate plant health. Improved contrast helps in identifying anomalies and changes in real time, facilitating better decision-making. Through continuous data updates and integrated analytics, the digital twin enhances situational awareness at both micro- and macro-scales. It empowers users to optimize resource usage (such as water, fertilizers, and pesticides), plan effective crop management strategies, and improve the resilience of

**Table 1** Experimental images for different image contrast enhancement techniques

| Serial no. | Original image | Image obtained by histogram equalization | Image obtained by intensity adjustment | Image obtained by adaptive histogram equalization |
|---|---|---|---|---|
| 1 |  | 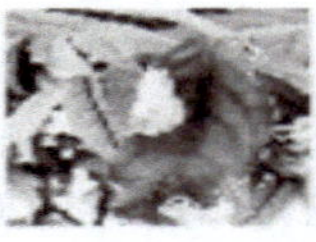 |  |  |
| 2 |  | 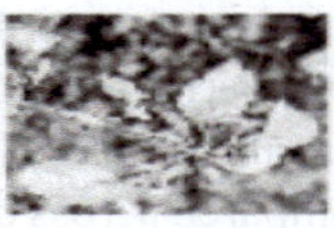 | 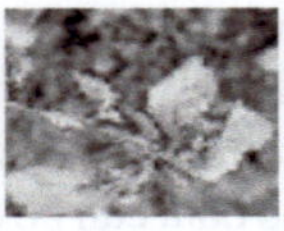 |  |
| 3 |  |  |  | 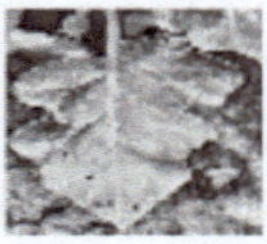 |
| 4 |  |  |  |  |
| 5 |  |  |  |  |
| 6 |  |  |  |  |
| 7 |  |  | 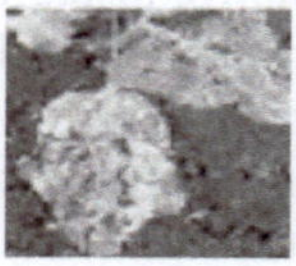 | 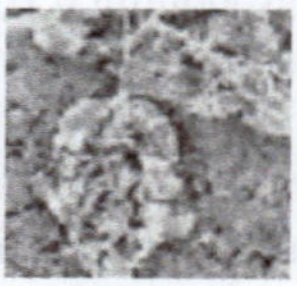 |
| 8 |  | 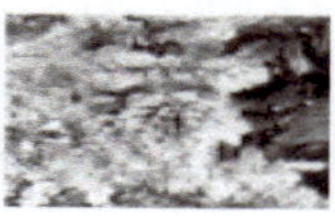 | 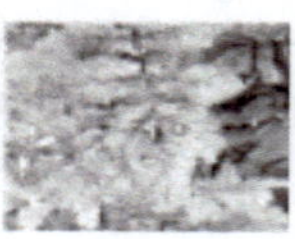 |  |

**Table 2** Entropy values of image obtained by using different image contrast enhancement

| Serial no. | Entropy of image obtained by histogram equalization | Entropy of image obtained by intensity adjustment | Entropy of image obtained by adaptive histogram equalization |
|---|---|---|---|
| 1 | 5.9232 | 6.8078 | 7.7032 |
| 2 | 5.9574 | 7.0748 | 7.0906 |
| 3 | 5.9810 | 7.6397 | 7.8943 |
| 4 | 5.9571 | 7.0978 | 7.6603 |
| 5 | 5.9648 | 7.2053 | 7.9477 |
| 6 | 5.9854 | 7.6436 | 7.9234 |
| 7 | 5.9108 | 7.0473 | 7.8187 |
| 8 | 5.9545 | 6.8141 | 7.8393 |

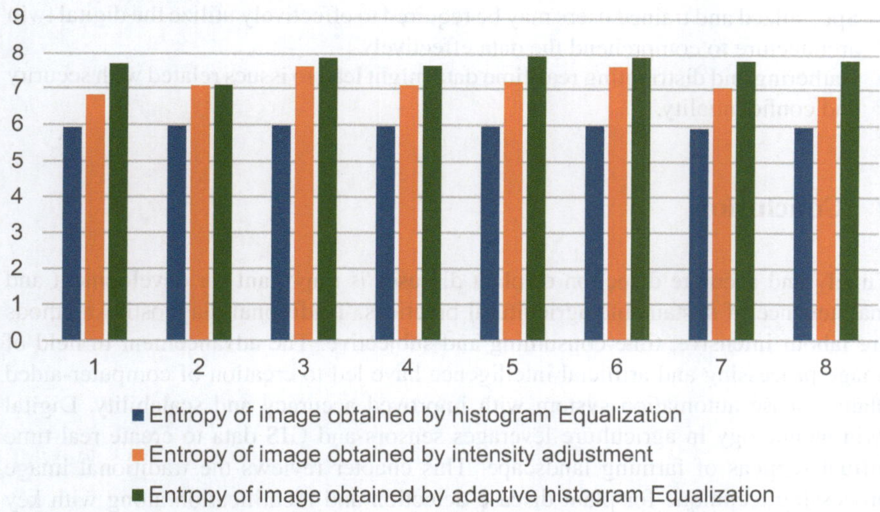


**Fig. 4** Comparison of entropy value of images obtained by different contrast enhancement method

agricultural systems. Ultimately, this framework not only improves operational efficiency but also contributes to long-term sustainable agricultural development and food security.

## 6.1 *Benefits of the Proposed Framework*

- A key benefit of this framework is enabling continuous monitoring of field conditions, providing real-time data integration.

- GIS maps facilitate spatially accurate and layered visualization, helping stakeholders understand complex agricultural environments.
- Provides farms with customized solutions that make the best use of resources like fertilizer and water according to the specific requirements.
- By visualizing and interacting with the environment, users can better comprehend spatial dynamics and relationships.

### 6.2 Disadvantages of the Proposed Framework

- The primary challenge is that the initial investment is very high and it may take a lot of financial resources to develop the technology and infrastructure.
- Managing data can be challenging since it requires to integrate and organize enormous quantities of data from various sources.
- Specialized and trained users may be required to effectively utilize the digital twin architecture to comprehend the data effectively.
- Gathering and distributing real-time data might lead to issues related with security and confidentiality.

## 7 Conclusion

Timely and accurate detection of plant diseases is important for development and maintenance of sustainable agricultural practices. Traditional diagnostics methods are labour intensive, time-consuming and subjective. The advancement in field of image processing and artificial intelligence have led to creation of computer-aided plant disease automation system with improved accuracy and scalability. Digital twin technology in agriculture leverages sensors and GIS data to create real-time virtual replicas of farming landscape. This chapter reviews the traditional image processing technique for plant disease detection and identification along with key components and methods used in digital twin environments to replicate physical agricultural system in their virtual counterpart. The AGIS-based framework which derives data from sensors, satellite imaging, drone in amalgamation of digital twin is also proposed in this chapter.

## References

1. Demilie WB (2024) Plant disease detection and classification techniques: a comparative study of the performances. J Big Data 11:5. https://doi.org/10.1186/s40537-023-00863-9
2. Ramanjot et al (2023) Plant disease detection and classification: a systematic literature review. Sensors. https://doi.org/10.3390/s23104769
3. Ebrahimi MA, Khoshtaghaza MH, Minaei S, Jamshidi B (2017) Vision-based pest detection based on SVM classification method. Comput Electron Agric 137:52–58

4. Li L, Zhang S, Wang B (2021) Plant disease detection and classification by deep learning—a review. IEEE Access 9:56683–56698. https://doi.org/10.1109/ACCESS.2021.3069646
5. Sengupta S (2022) Role of image processing in artificial intelligence and Internet of Things. Comput Vis Internet Things 1:11
6. Sengupta S, Mittal N, Modi M (2018) A survey of techniques used in processing and mining of medical images. In: Panda B, Sharma S, Roy N (eds) Data science and analytics. REDSET 2017. Communications in computer and information science
7. Kim J-S, Chung S, Ko M, Song J, Shin S (2025) Comparison of image preprocessing strategies for convolutional neural network-based growth stage classification of butterhead lettuce in industrial plant factories. Appl Sci 15:6278. https://doi.org/10.3390/app15116278
8. Tariku G, Ghiglieno I, Simonetto A, Gentilin F, Armiraglio S, Gilioli G, Serina I (2024) Advanced image preprocessing and integrated modeling for UAV plant image classification. Drones. 8:645. https://doi.org/10.3390/drones8110645
9. Zhang Y-C, Mao H-P, Hu B, Li M-X (2007) Feature selection of cotton disease leaves image based on fuzzy feature selection techniques. In: Proceedings of the 2007 international conference on wavelet analysis and pattern recognition, Beijing, China, November 2007, pp 124–129
10. Yang W, Yuan Y, Zhang D, Zheng L, Nie F (2024) An effective image classification method for plant diseases with improved channel attention mechanism aECAnet based on deep learning. Symmetry 16(4):451. https://doi.org/10.3390/sym1604045
11. Saad MH, Salman AE (2024) A plant disease classification using one-shot learning technique with field images. Multimed Tools Appl 83:58935–58960. https://doi.org/10.1007/s11042-023-17830-4
12. Shafik W, Tufail A, De Silva Liyanage C et al (2024) Using transfer learning-based plant disease classification and detection for sustainable agriculture. BMC Plant Biol 24:136
13. Gupta M, Kumari M, Ved C (2021) Digital twin in agriculture sector: detection of disease using deep learning. In: Digital twin technology. CRC Press, pp 47–60
14. Dai M, Shen Y, Li X, Liu J, Zhang S, Miao H (2024) Digital twin system of pest management driven by data and model fusion. Agriculture 14(7):1099. https://doi.org/10.3390/agriculture14071099
15. Abayadeera MR, Ganegoda G (2024) Digital twin technology: a comprehensive review. Int J Sci Res Eng Trends 10:1485–1504. https://doi.org/10.61137/ijsret.vol.10.issue4.199
16. Sharma L, Garg PK (eds) (2025) Digital twins and simulation technology concepts and applications, 1st edn. Chapman and Hall/CRC. https://doi.org/10.1201/9781003582489
17. Sharma L, Garg PK (eds) (2024). Deep learning in Internet of Things for next generation healthcare, 1st edn. Chapman and Hall/CRC. https://doi.org/10.1201/9781003451846
18. Sharma L, Garg PK (eds) (2023) Technological prospects and social applications of society 5.0, 1st edn. Chapman and Hall/CRC. https://doi.org/10.1201/9781003324720
19. Sharma L, Carpenter M (eds) (2022) Computer vision and Internet of Things: technologies and applications, 1st edn. Chapman and Hall/CRC. https://doi.org/10.1201/9781003244165
20. Sharma L, Garg PK (eds) (2021) Artificial intelligence: technologies, applications, and challenges, 1st edn. Chapman and Hall/CRC. https://doi.org/10.1201/9781003140351
21. Sharma L (ed) (2020) Towards smart world: homes to cities using Internet of Things, 1st edn. Chapman and Hall/CRC. https://doi.org/10.1201/9781003056751
22. Sharma L, Garg PK (eds) (2019) From visual surveillance to Internet of Things: technology and applications, 1st edn. Chapman and Hall/CRC. https://doi.org/10.1201/9780429297922
23. Sharma L (2019) Object detection with background subtraction. LAP LAMBERT Academic Publishing, SIA OmniScriptum Publishing, Latvia, European Union, February 2019. ISBN: 978-613-7-34386-9
24. Borra S, Thanki R, Dey N (2019) Satellite image enhancement and analysis. In: Satellite image analysis: clustering and classification. Springer Singapore, Singapore, pp 13–29
25. Mridha MF, Dey N (2025) Machine vision in plant leaf disease detection for sustainable agriculture

26. AlShahrani AM, Al-Abadi MA, Al-Malki AS, Ashour AS, Dey N (2018) Automated system for crops recognition and classification. In: Computer vision: concepts, methodologies, tools, and applications. IGI Global Scientific Publishing, pp 1208–1223
27. Rajinikanth V, Dey N, Satapathy SC, Kamalanand K (2019) Inspection of crop-weed image database using kapur's entropy and spider monkey optimization. In: Soft computing for problem solving: SocProS 2018, vol 1. Springer Singapore, Singapore, pp 405–414
28. Tharwat A, Gaber T, Awad YM, Dey N, Hassanien AE (2015) Plants identification using feature fusion technique and bagging classifier. In: The 1st international conference on advanced intelligent system and informatics (AISI2015), November 28–30, Beni Suef, Egypt. Springer International Publishing, Cham, pp 461–471
29. Chaki J, Dey N, Moraru L, Shi F (2019) Fragmented plant leaf recognition: Bag-of-features, fuzzy-color and edge-texture histogram descriptors with multi-layer perceptron. Optik 181:639–650

# Chapter 12
# The Role of Artificial Intelligence (AI) and Geographic Information Systems (GIS) in Powering Digital Twins for Digital Plant Pathology

**Rene Heim, Pradeep Kumar Garg, Prince Kumar, Davide Mattioli, and Martin Kappas**

**Abstract** Agriculture is increasingly challenged by human-induced climate change, land degradation and increasing crop losses caused by pests and pathogens. Reductions in crop yields of 17–30% across major crops' unevenly affect food insecure regions. Yet while digital tools from basic farm-management apps to sophisticated automation have been spreading, cost, connectivity and expertise remain significant barriers preventing such technology from becoming more widespread. Incorporating such technologies into full digital twins has the potential to link disparate data streams, improve decision-making reliability and deliver interventions that are specifically relevant to a locale. Yet most agricultural digital twins are still confined to experimental scale due to the difficulty of modelling life, infrastructure deficits in rural areas and difficulties in ensuring robustness of the model under environmental diversity. These gaps will be filled with the development of standardized protocols for data, open access archives and repositories, large-scale interdisciplinary collaboration, and focused training efforts. As governments and innovators seek to reconcile yield with sustainability, digital twins might offer a way to truly precise, adaptive and resilient farm management provided the technical, economic and policy challenges are addressed in concert.

**Keywords** AI (artificial intelligence) · Crop disease detection · Digital twins · Drones for agriculture · Geographic information systems (GIS) · Internet of Things (IoT) in agriculture · Machine learning (ML) · Remote sensing · Uncrewed aerial systems (UAS) · Vegetation indices

---

R. Heim · D. Mattioli · M. Kappas (✉)
GIS and Remote Sensing Section, Institute of Geography, Georg-August University Goettingen, Goettingen, Germany
e-mail: mkappas@gwdg.de

P. K. Garg · P. Kumar
Indian Institute of Technology, Roorkee, India

 
L. Sharma et al. (eds.), *Digital Twin Technology in Geospatial Science: Trends, Applications and Challenges*, Data-Intensive Research,
https://doi.org/10.1007/978-3-032-20311-3_12

## 1 Introduction

Agrifood systems still provide the food, employment and economic growth of billions of people and are a cornerstone of many national economies, particularly across the Global South [10]. Yet contemporary farming now faces a convergence of threats: intensifying climatic change, growing freshwater scarcity, declining soil health and fertility, and the rapid spread of invasive or fast-evolving pests and pathogens [16, 36, 42]. International frameworks increasingly recognize that plant health is fundamental to food security, trade, and environmental sustainability. The combination of global standards (e.g., International Plant Protection Convention), regional policies (e.g., EU Common Agricultural Policy, EU Green Deal), and climate commitments (e.g., UNFCCC Paris Agreement) creates a comprehensive but challenging policy landscape. A recent global synthesis across wheat, rice, maize, potato, and soybean estimated that diseases and animal pests erase 17–30% of attainable yields each season, with the heaviest burdens (>30%) falling on food-insecure regions in sub-Saharan Africa and South Asia [42]. These losses translate into billions of dollars in lost revenue, lower grain quality, and heightened nutrition insecurity. Most farmers, especially those in developing nations, do not have easy access to smart and flexible tools to detect and manage biotic (e.g., diseases) and abiotic (e.g., drought) crops stress.

Smart agriculture, and related concepts such as "precision farming" and "Agriculture 4.0" (see Aggarwal and Verma [2] for an evolution from Agriculture 1.0 to Agriculture 4.0), is aiming at transforming food production by combining digital technologies, such as advanced sensors and data analytics, into every stage of the farming cycle, from variety selection to post-harvest handling [51–53]. In Brazil, a nationwide survey of 504 farmers showed that 84% already use at least one digital tool. They mention primarily *internet connectivity*, *mobile applications* and *software* for basic farm management. Higher-complexity technologies such as *variable-rate machinery*, *automation* and *digital mapping* remain less common. Perceived benefits centre on higher productivity, easier marketing and better planning, yet adoption is still constrained by high acquisition costs, limited rural connectivity and a shortage of specialised service providers [6]. Data gathered from various sources, such as field sensors, remote sensing imagery, UAS (uncrewed aerial systems), soil probes, and weather stations, can provide valuable information, such as humidity, pH value, temperature, soil nutrient content and moisture, disease and pest scores, soil depletion levels, spatial heterogeneity and multiple other environmental factors relevant for productivity agriculture. The data collected is to be seamlessly integrated to get accurate and reliable results. For precise and reliable data collection, sensors and equipment must be regularly calibrated. The huge data collected may not be always required, and only relevant data should be selected for empirical and deterministic models to evaluate the diverse factors and their impact on crop production and stress.

Across Europe, a study of more than 1800 arable and 1300 livestock farms in Bavaria found that automation and decision-support solutions are spreading, though overall digitalisation is modest. Section control, variable-rate application and barn robotics are expected to reach adoption rates of 15–20% within five years, but entry technologies such as automatic milking systems, digital field records and automatic steering strongly influence subsequent uptake of additional tools. In both crop and livestock systems, farmers give priority to user-friendly automation that reduces labour and workload [13]. In the United States, digital transformation is promoted as a pathway to reconcile higher yields with environmental sustainability. Site-specific irrigation scheduling, sensor-guided fertiliser placement and robotic cover-crop planting are among the technologies discussed for tackling water scarcity, nitrogen losses and herbicide resistance. Adoption of sustainable practices depends on money, habits, social influence, and trust in data.

Natural habitats in less intensively farmed areas can ease some of the harm from climate change and intensive agriculture but cannot prevent it [24]. Asia presents a contrasting picture. Rapid progress in China, Japan and South Korea is driven by strong governmental support, AI capabilities and investment in internet of things (IoT) infrastructure, whereas much of Southeast and South Asia still relies on traditional practices. Reviews stress the urgency of smart agriculture in the face of climate change, labour shortages and water scarcity, and call for stronger policies on data security, risk prevention and standardization [48, 54, 55].

Currently, more digital tools (e.g., weeding robots, crop spraying drones) are being tested and integrated within the context of agriculture. Yet, persistent challenges in data quality, connectivity, interoperability and economic viability show that these technologies are not yet universally accessible or reliable. A major barrier is the integration of multi-modal data into integrated models that have a direct impact on farm management. Thus, digital twins improve decision-making process by facilitating more precise, adaptive and sustainable farm management [38, 39].

## 2 Digital Twins in Agriculture

The digital twin is a digital representation of a real-world object that may be used to remotely monitor the real-world object's performance, and simulate real-world scenarios and their outcomes [59–63]. The digital replica of a farmland might reflect its characteristics, and spatial heterogeneity, like field size, filed shape, slope, soil type, micro-climate, sunlight exposure, cropping patterns, irrigation, etc. [7]. The conceptual framework for agriculture digital twins consists of the following four phases: (1) data collection, (2) data processing, (3) data analysis and evaluation, and (4) management response [7, 38].

Despite their potential, digital twins in agriculture remain largely in the experimental stages. Most current implementations are focused on specific, narrow applications rather than comprehensive farm-wide systems, such as monitoring individual crops or optimizing irrigation, or tracking animals in controlled environments and greenhouses. There have been notable examples such as the FIWARE Accelerator Programme in Europe [40], or the Fraunhofer COGNAC lighthouse project (https://www.iese.fraunhofer.de/en/project/cognitive-agriculture.html). The latter has been developing a secure, interconnected "agricultural data space" to integrate advanced sensor, automation, and robotics solutions. Such collaborative approaches are likely crucial to co-designing functional digital twins, an approach in which multidisciplinary stakeholders (e.g., farmers, scientists, policymakers) are treated as equal collaborators working toward common goals. While we are waiting for successful collaborative efforts, current systems often lack the sophisticated integration and real-time feedback communication (Fig. 1) that defines true digital twins. Instead, current systems are functioning more as "digital shadows" that primarily collect and visualize data without the provision of comprehensive predictive modelling or autonomous control capabilities. The agricultural community's adoption has been slower compared to manufacturing and other industries, partly due to the complexity of modelling living biological systems [7, 38, 39].

Infrastructure constraints, including intermittent internet connectivity, unreliable power supply and the harsh climate conditions that could destroy vulnerable sensors [25, 38] are also hurdles to develop operational digital twins in farming. Apart from sensor integration, the robustness of predictive models represents a basic aspect for the credibility of digital twins. Particularly models which aim to provide a close description of complex vegetated systems, such as those present in agriculture, can be plagued by performance decay due to data drift caused by dynamic changes to biological processes and weather patterns (climate variability), as well as spatial and

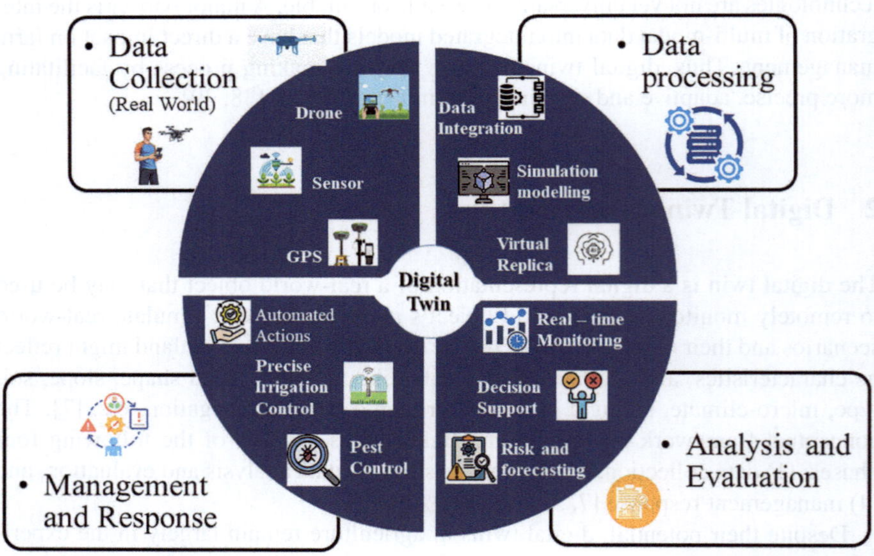


**Fig. 1** Conceptual framework of a digital twin in agriculture. Sensor data (e.g., remote sensing, GPS navigation, on-farm sensors) collected from the real environment is processed in a digital space, enabling continuous feedback and informed management decisions. This integrated approach links real-world observation with virtual modelling to optimize agricultural practices and support precise, data-driven interventions. *Image designed by Rene Heim. Contains icons designed by* Kanin Abhiromsawat via Vecteezy.com

temporal heterogeneity within agricultural settings that would require continuous calibration and validation effort across various agriculture contexts [44]. It will take collective action to overcome these barriers. But digital twin tech is competing to become the key agent of change towards sustainability in agriculture. And finding workable solutions has become imperative as demand for food grows within a fast-expanding global population.

## 3 Geographic Information Systems and Remote Sensing in Crop Disease Monitoring

To address a significant challenge in the integration of digital twins, geographic information systems (GIS) can provide foundational layers for identifying local and landscape heterogeneity and visualizing spatial-based crop health data. GIS provides the spatial framework for organization, analysis and display of data around specific locations, which is vital given the inherently spatially variable nature of farming [14, 23, 29]. Soil moisture, vegetation index data, rainfall in the specific geographic area and even pest and disease incidence is most useful when it occurs within a geospatial framework. For example, a wheat field might have different fertility and moisture at the beginning to the end. One part of the field may be prone to fungal infections, due to shaded places and proper microclimates from leave and canopy structure. Another part of the field may suffer from nutrients as large farm machinery causes compaction, limiting soil porosity, water flow and drainage, root growth and the ability of plants to absorb water and nutrients.

Digital plant pathology relies on understanding that plant diseases can bring measurable changes in optical and structural properties that can be identified using electrochemical and optical sensing technologies. Spectral reflectance signatures can indicate disease-related physiological and biochemical changes, while LiDAR-derived 3D point clouds can detect structural canopy symptoms, like changes in leaf angle or leaf area [56–58].

The most accessible entry point into optical sensing for plant health monitoring involves standard RGB (red–green–blue) cameras. RGB sensors are highly affordable and portable, and their ability to capture images at very high spatial resolution makes them ideal for detailed visual analysis and monitoring of small disease symptoms at the leaf-scale over large areas. However, because they operate in only three broad spectral bands (red, green, and blue), they are limited in their ability to detect subtle physiological changes in plants that are often associated with bands beyond the visible RGB region (e.g., near-infrared or shortwave-infrared) provided by multispectral or hyperspectral sensors. Deployed with uncrewed aerial systems, RGB cameras have demonstrated potential for disease detection, for field-scale monitoring [31]. The technology becomes particularly powerful when combined with machine learning approaches for image analysis and disease classification [3, 18, 21].

Moving beyond basic RGB imaging, multispectral and hyperspectral sensors may enhance capabilities for disease detection if the data is well pre-processed and interpreted (e.g., camera calibration, lens distortion correction and radiometric calibration). Radiometric calibration plays a significant role for multispectral and hyperspectral data and correct signal interpretation especially for time series analysis [1, 4, 11]. Multispectral imaging systems operate across multiple discrete, usually broad (~20–50 nm), spectral bands. In contrast to RGB cameras, these systems operate in non-visible regions in addition to the visible range (NIR and SWIR), providing information of biochemical and structural properties of plants that are invisible to humans. Hyperspectral imaging also provides potential as an optical remote sensing for plant health assessment and is underutilized in such complex field environments with its applications more commonly used for controlled indoor scenarios at present [29, 30, 35]. This low adoption in the field can be attributed to high costs and complexity associated with hyperspectral systems, as well as their large computational requirements for storage and processing. The predominance of line-scanning sensor designs has made imaging prone to motion artifacts, geometric distortion, and alignment errors [1]. These practical hurdles, including cost, operational complexity, data processing burden, and susceptibility to artefacts, further explain why hyperspectral imaging has not yet achieved widespread adoption for large-scale or structurally complex outdoor plant disease monitoring. Hyperspectral sensors are typically designed to capture spectral signals across much wider ranges in narrow, contiguous bands, covering ultraviolet (250–400 nm), visible (400–700 nm), near-infrared (700–1300 nm), and shortwave infrared (1300–2500 nm) wavelengths. At the landscape scale, a satellite-based hyperspectral sensor has been available since 2022. The environmental mapping and analysis program (EnMAP) is a German hyperspectral satellite mission that monitors and characterizes the Earth's environment at a global scale. EnMAP measures geochemical, biochemical and biophysical variables providing information on the status and evolution of terrestrial and aquatic ecosystems in a resolution of $30 \times 30$ m.

Several complementary sensing technologies can enhance disease detection and monitoring. Thermal infrared (TIR) imaging captures radiation, in the long-infrared thermal range (8–14 μm), providing information about plant temperature and transpiration patterns [4]. Plant–pathogen interactions often disrupt stomatal function, producing temperature changes in affected leaf areas that can be detected with thermal infrared (TIR) imaging. While TIR has the potential to allow pre-symptomatic detection in pathosystems, the accuracy of these measurements can be compromised by sensor signal drift, particularly due to wind-induced cooling of the camera or direct solar heating of the sensor. Such drift, combined with confounding effects from abiotic stress, can significantly reduce reliability in field conditions. [5, 31]. Still, it has been successfully integrated into multi-sensor approaches to provide robust detection of biotic plant stress of *Xylella fastidiosa* [49]. Further, light detection and ranging (LiDAR), while primarily used for structural measurements, generates dense 3D point clouds that can reveal disease-related changes in plant architecture such as leaf area and leaf angle. Since such structural and geometric plant traits can be affected by plant diseases, LiDAR and other concepts generating 3D point clouds (e.g., stereo vision cameras) can provide complementary information to spectral approaches [22].

One of the most significant contributions of remote sensing and GIS to plant pathology lies in its ability to reveal and quantify spatial heterogeneity in disease patterns [19]. Spatial heterogeneity refers to the complex distribution of land cover types across landscapes and the spatial patterns or frequency of those land cover types. Even at field scale, spatial heterogeneity is present and affects the dynamics of crops and soil [8, 15, 34, 37]. As demonstrated by [12], landscape heterogeneity is inherently scale-dependent, meaning patterns change with spatial resolution. The problem of scale has been a long-lasting challenge in the interpretation of remote sensing data [47].

Moreover, the objectives of sensor-based disease detection and monitoring needs to be set in parallel with the management conditions actually practiced, if sensing outputs are to be translated into an actionable intervention. For instance, early plant disease detection is a relative concept as the degree of symptom development and incubation periods can greatly differ between various pathosystems. That is a testament to the range of biology around host–pathogen interactions and how you deal with them. "Early" can refer to pre-symptomatic detection, identifying pathogens before visible symptoms appear, this period that varies from few hours to weeks depending on the incubation phase of the pathogen. Additionally, early can be defined as detecting disease when there are no visible symptoms or at an earlier stage of the damage and intervention is provided sooner. Accordingly, prevention or monitoring is to be coordinated with the management, and "early" detection itself is of value only if accompanied by some form of intervention.

## 4 Role of Artificial Intelligence (AI) and Machine Learning in Crop Disease Detection

Detection and prediction of crop diseases is important for maximising the yield and minimising the pesticide applications. Machine learning can be used to develop automated disease detection based on satellite remote sensing data, UAV data, and ground-based sensors. By learning complex patterns in multi-band imagery, machine learning models can identify subtle spectral and textural changes in plant canopies [43].

Early applications of artificial neural networks for plant disease assessments were published by [23, 28]. They developed a monitoring concept for plant health using a combination of neural networks and statistical classifiers and identified nutrient deficiencies in lettuce leaves using a video camera system where analogue images first had to be digitized for processing. The demand for architectures being able to recognize local patterns led to further developments of convolutional neural networks (CNNs) for disease detection [28, 46]. It is now common practice to train CNNs on cleaned leaf images (e.g., The Plant Village Dataset) to show that accurate class predictions are possible across a broad range of diseases and plant species. However, real-world field imagery still presents challenges as overlapping leaves, variable illumination, and heterogeneous backgrounds, demand more robust models and pre-processing. Mask R-CNN instance-segmentation networks, for example, have been adapted to extract individual leaf instances from UAV-acquired multispectral ortho-mosaics, enabling leaf-based disease parameters to be computed with high precision even under disease pressure and changing light conditions [3].

Vision transformers (ViTs) represent the next evolution in deep architectures, leveraging self-attention mechanisms to capture long-range dependencies in image data. SugarViT, a multi-objective ViT model trained with deep label distribution learning, has shown excellent performance in scoring Cercospora leaf spot severity in sugar beet directly from UAV multispectral imagery [18]. By modelling severity as a probability distribution rather than a discrete label, the approach allows for uncertainty in expert annotation and provides interpretable confidence values. In addition, training on environmental data (e.g., growing degree days, accumulated infection cycles) additionally improves model robustness when tested upon disease severity scores.

A continuing challenge is how to adapt fundamental models, including ImageNet, to specific domain like GIS-aided plant disease monitoring. While these models design powerful general-purpose feature representations, their performance tends to decrease when applied to domain-specific tasks where fine grained spectral-textural clues or spatial context from geospatial variables, such as elevation, soil moisture, and microclimate patterns, are required. A key example is that using a geospatial foundation model encoder and part down-sampling with a frozen decoder, the model in its foundation form can estimate biomass across diverse Brazilian ecosystems [33], as accurately as another baseline method called the U-Net approach, which requires thirteen times more trainable variables. Although this method saves on computational time, it could limit flexibility in more complicated and changing environments.

ML models need to generalize across different climatic zones, sensor types, and management practices, requiring large, diverse, and accurately labelled datasets. Data-centric techniques, such as confidence-driven label noise reduction, relevance-based sample weighting to counter geographic domain shift, and slice discovery to reveal model weaknesses on specific subpopulations, are essential for curating robust training and evaluation sets [41]. Moreover, interpretability and explainability of AI predictions are critical for fostering end-user trust in automated disease alerts [43]. In the literature [29, 31, 41, 43] a convergent set of hurdles emerges: optical signatures of disease are still poorly generalised because current training data lack diversity, accuracy, consistency, unbiasedness and task-specific relevance, while reference labels remain scarce, noisy and unevenly distributed across species, pathogens, climates and phenological stages; sensor outputs are sensitive to illumination, viewing geometry and atmospheric effects, yet calibration standards, metadata protocols and shared spectral libraries are fragmentary, impeding reproducibility [17, 45], data fusion and transferability from leaf to canopy, greenhouse to field, and field to satellite scales, confounding abiotic stresses, mixed infections and cultivar-dependent responses blur spectral signals, so detection pipelines must incorporate epidemiological context, uncertainty propagation and mechanistic insight rather than rely on black-box pattern recognition. Seeleyet al. [44], in a natural environment context, specifically asked why *models go wrong* and found that spectral classification models often fail when applied across diverse regions because environmental dissimilarity, particularly differences in climate variables like temperature and precipitation, cause spatially structured spectral phenotypes that deviate from training data, rather than mere geographic distance. Just as wild plant populations exhibit spatially structured spectral phenotypes due to local adaptation and environmental pressures, agricultural genotypes planted in different environments experience similar environmental distance effects that alter their spectral signatures [9]. Even genetically uniform crops exhibit variable spectral responses due to functional and structural trait plasticity triggered by local soil conditions, climate variations, and management practices. This genotype-specific environmental sensitivity creates additional complexity for spectral models, as the same environmental distance may produce different magnitudes of spectral change depending on the genetic background. This emphasizes that spectral models in agriculture are vulnerable to the same "environmental distance" effects identified by [44]. However, most models treat data as static benchmarks instead of dynamic systems, meaning feedback loops for active learning, quality auditing and bias mitigation are missing, and sparse ground truth, privacy constraints and incompatible data policies further inhibit cross-regional deployment,finally, the socio-technical chasm between sensor engineers, plant pathologists, modellers and end-users slows translation into actionable decision support, especially when regulatory frameworks for site-specific interventions or automated spot-spraying remain unresolved. Addressing these challenges requires coordinated, data-centric curation, interoperable standards, open repositories that couple optical traits with epidemiological trajectories, and hybrid workflows in which remote sensing continuously updates and is updated by process-based disease simulators; only then can the high-frequency, multi-scale data streams needed to populate, calibrate and validate crop-scale digital twins be realised, allowing virtual replicas of plant–pathogen–environment systems to guide real-time disease management and resource use.

The synergy between AI and GIS within a digital twin framework creates a proactive, data-driven system theoretically capable of monitoring crop health, forecasting disease outbreaks, and delivering targeted alerts. For example, deep learning models trained on multispectral and hyperspectral images were able to detect early-stage infections of diseases, such as wheat rust, rice blast, or maize leaf blight [32]. These detections, when mapped and visualized using a GIS, enable spatial tracking of disease spread and risk zoning, facilitating timely and site-specific interventions. Digital twins are desirable as they would not only safeguard yields but also optimize resource utilization by minimizing unnecessary pesticide application and reducing environmental impact.

At the core of digital twins [50] lies the training of AI models aiming to detect disease signatures from images and other multimodal sensor data. Various modelling techniques are currently recognized as "state-of-the art" and used based on the available data type and task complexity.

# 5 Model Types

## 5.1 Classification Models (Image-Based)

- Convolutional Neural Networks (CNNs): Primary architecture for recognizing disease patterns in image data.
- Popular CNN Variants: VGG16, ResNet, and InceptionV3.
- Transfer Learning: Pre-trained models (on ImageNet or PlantVillage datasets) are fine-tuned on agricultural data for enhanced performance with limited training samples.

*Object Detection and Segmentation*

- YOLO (You Only Look Once) and Faster R-CNN: For detecting and localizing disease spots in images.
- U-Net and Mask R-CNN: For pixel-level disease segmentation.

*Time-Series and Forecasting Models*

- LSTM (Long Short-Term Memory) networks: For modeling disease progression based on NDVI trends, temperature, and rainfall.
- GRU (Gated Recurrent Units): For lightweight time-series analysis in real-time applications.

*Ensemble and Hybrid Models*

- Multi-classifier approaches: Combine multiple classifiers (e.g., random forests + CNNs) for enhanced robustness.
- Multi-input architectures: Process both images and sensor data simultaneously as model inputs.

*Spatial Machine Learning*

- Spatial autocorrelation-based models: Including spatial regression and geostatistics to enhance disease clustering and spread modeling.

# 6 Data Types

## 6.1 Multi-band Drone Imagery

Multi-band imagery consists of images captured by UASs equipped with RGB (red, green, blue) or multispectral (RGB + ear infrared, and red edge) cameras that measure reflectance in different spectral bands. These images are analysed using CNNs to detect spatial patterns of plant diseases, and distinguish between healthy and diseased vegetation. These images can be used to derive vegetation indices, like NDVI (normalized difference vegetation index), NDRE (normalized difference red edge), and GNDVI (green normalized difference vegetation index), which may be used as input to machine learning algorithms for disease assessment. The high-resolution images provide detailed information on training data required for supervised learning algorithms. The classification can be supported by ground truth validation. The time-series data of UAV images collected by different flights enables the identification of temporal patterns in diseases.

# 7 Satellite-Derived Vegetation Indices

Remote sensing based vegetation indices provide a good source of information for agricultural disease identification. Sentinel-2, Landsat-8/9, MODIS, EnMAP, and PlanetScope can provide multispectral and hyperspectral images for vegetation health indicators. These images can be used to derive vegetation indices that indicate plant health. The NDVI provides basic information about chlorophyll and biomass levels provided it is calibrated with the cropping systems, whereas the enhanced vegetation index (EVI) gives improved accuracy by minimising the effect of atmosphere and noise of soil background. The soil adjusted vegetation index (SAVI) offers useful information for sparse vegetation cover regions, and the normalized difference water index (NDWI) can track plant water contents, which distinguishes between water stress and disease-related stresses.

The use of NDVI in agriculture can sometimes lead to misinterpretations of vegetation health. It can only provide clue to several vegetation characteristics, including photosynthetic activity, leaf vitality, vegetation density, and canopy structure. The widespread use of NDVI in remote sensing applications originated from its simplicity which has sometimes masked NDVI's limitations, leading to those applications where other specialized indices give more accurate assessments.

For agricultural disease detection specifically, relying solely on NDVI presents several significant challenges that can compromise monitoring accuracy. Disease stress may not always reduce chlorophyll content at the early disease stages to be detectable above the background noise created by soil reflectance and atmospheric interference. Additionally, saturation effects in NDVI calculations of dense vegetation may mask subtle disease-related changes in vegetation health. Perhaps most importantly, different diseases affect plant reflectance in different ways across various spectral regions. However, the NDVI is limited to the red and near-infrared region of the electromagnetic spectrum. Consequently, the field would likely benefit from moving toward multi-index approaches that combine NDVI with more specialized vegetation indices or even combined with mechanistic modeling approaches.

Still, satellite-derived imagery offers several advantages for agricultural monitoring. It provides consistent, large-scale coverage ranging from individual villages, including small-scale farms, to entire nations across vast agricultural regions. Much of the satellite data is open and free (programs like USGS, Copernicus, NASA) which allows access for all to advanced agricultural monitoring worldwide. The availability of historical time-series data spanning multiple years enables sophisticated modeling of disease progression patterns across different seasons and climatic conditions. For instance, the high-resolution vegetation phenology and productivity (HR-VPP) service from Copernicus now delivers consistent, high-resolution (10–20 m) maps of key vegetation metrics, such as start-of-season, end-of-season, season length, peak greenness, and aggregated productivity indices (LAI, fAPAR, NDVI). Updated at ten-day or monthly intervals, HR-VPP enables precise monitoring of crop and ecosystem phenology, early detection of anomalies, and accurate assessment of biomass growth and carbon uptake at field to regional scales, supporting agricultural management, biodiversity monitoring, and policy reporting.

In disease detection models, satellite-derived vegetation indices serve multiple analytical functions. Weekly or bi-weekly vegetation index data can feed into long short-term memory (LSTM) networks and other recurrent neural networks (RNN) for disease outbreak forecasting, leveraging temporal patterns to predict future disease occurrences. For classification, mean NDVI values computed over specific areas provide important input to machine learning models (i.e., random forests or support vector machines). This information can be integrated in geographic information systems (GIS) to create disease spread risk maps, thus providing distribution patterns of disease across agricultural lands.

## 7.1 IoT Sensor Data

IoT sensor networks is one of the essential components which is widely utilized to set up plants regarding precision agriculture. The sensors can monitor the micro-environmental change that may cause diseases in agriculture fields in realtime. These sensors check the soil's moisture, temperature, pH, and electrical conductivity to guess what the root zone is like. Climate sensors measure ambient temperature, relative humidity, rainfall, solar radiation, and leaf wetness duration, which can reflect enough moisture surrounding characteristics for pathogen growth. In this scenario, the entire sensor network typically takes measurements each 15 min or less and reports that data to a cloud-based platform for real-time analysis.

IoT sensors ought to integrate with disease detection models, and their data should be used for multiple types of analysis to enhance prediction accuracy and response time. These sensors collect information about the environment that goes straight into AI models as feature inputs for predicting disease risk scores, this gives the models the context they need to make accurate predictions about the likelihood of disease. Also, the systems will make threshold-based directives to notify a disease threat or condition. Specifically, under conditions like downy mildew risk, alerts are raised when humidity levels exceed 90% and leaf wetness persists for prolonged durations. IoT sensors send out streams of data all the time, these streams are perfect for time-series modeling methods like ARIMA, Prophet, or LSTM, which can find patterns over time and use them to predict future disease risk based on changes in the environment.. Finally, the ability to deploy of edge AI (i.e., artificial intelligence processing data locally rather than in remote cloud servers) allows for real-time inferences on-site using lightweight models running on microcontrollers like NVIDIA Jetson or Raspberry Pi devices. This means that decisions can be made and responses can be given right away without needing to connect to the cloud.

# 8 Conclusion

The development of digital agriculture has advanced rapidly, fueled by advances in sensing, data analytics, and automation. Although this progress has yet to reach the maximum potential for digital twins, fully functioning virtual counterparts that continuously reflect and predict field dynamics, remains nascent. Key impediments include (1) the need for standardized approaches to capture and integrate multi-source data; (2) the lack of high-quality, interoperable datasets that cover a wide range of crops, climates, and management practices; (3) he technical limitations of rural infrastructure, such as unreliable power and connectivity; and (4) the inherent challenges of modelling biological processes and making sure that models are valid in changing environmental conditions. Coordinated action is necessary to overcome these obstacles. First, open and FAIR-compliant data platforms must be established. Second, resilient models with good explanatory capabilities should be developed, including those that adjust to data drift. Third, interdisciplinary collaboration among agronomists, engineers, modellers, and policymakers should be expanded. Finally, farmers and consultants must be trained to in concert, these measures will enable digital twins to evolve from basic research prototypes to practical decision-support algorithms, deploying site-specific irrigation, nutrient management, pest and disease prediction, and completely autonomous field activities. Integral to balancing the need to feed the population and safeguarded ecosystems, they will guide agriculture to a more accurate, robust, and sustainable future.

# References


1. Aasen H, Honkavaara E, Lucieer A, Zarco-Tejada PJ (2018) Quantitative remote sensing at ultra-high resolution with UAV spectroscopy: a review of sensor technology, measurement procedures, and data correction workflows. Remote Sens 10:1091. https://doi.org/10.3390/rs10071091
2. Aggarwal S, Verma A (2022) Transformations in the ways of improving from agriculture 1.0 to 4.0. In: 2022 5th international conference on contemporary computing and informatics (IC3I), pp 170–174. https://doi.org/10.1109/IC3I56241.2022.10072298
3. AlShahrani AM, Al-Abadi MA, Al-Malki AS, Ashour AS, Dey N (2018) Automated system for crops recognition and classification. In: Computer vision: concepts, methodologies, tools, and applications. IGI Global Scientific Publishing, pp 1208–1223
4. Barreto A, Reifenrath L, Vogg R, Sinz F, Mahlein A-K (2023) Data augmentation for mask-based leaf segmentation of UAV-images as a basis to extract leaf-based phenotyping parameters. KI—Künstliche Intelligenz 37:143–156. https://doi.org/10.1007/s13218-023-00815-8
5. Berger K, Machwitz M, Kycko M, Kefauver SC, Van Wittenberghe S, Gerhards M, Verrelst J, Atzberger C, Van Der Tol C, Damm A et al (2022) Multi-sensor spectral synergies for crop stress detection and monitoring in the optical domain: a review. Remote Sens Environ 280:113198. https://doi.org/10.1016/j.rse.2022.113198
6. Bock CH, Chiang K-S, Del Ponte EM (2021) Plant disease severity estimated visually: a century of research, best practices, and opportunities for improving methods and practices to maximize accuracy. Trop Plant Pathol 47:25–42. https://doi.org/10.1007/s40858-021-00439-z
7. Bolfe ÉL, Jorge LA de C, Sanches ID, Luchiari Júnior A, da Costa CC, Victoria D de C, Inamasu RY, Grego CR, Ferreira VR, Ramirez AR (2020) Precision and digital agriculture: adoption of technologies and perception of Brazilian farmers. Agriculture 10:653. https://doi.org/10.3390/agriculture10120653
8. Cesco S, Sambo P, Borin M, Basso B, Orzes G, Mazzetto F (2023) Smart agriculture and digital twins: applications and challenges in a vision of sustainability. Eur J Agron 146:126809. https://doi.org/10.1016/j.eja.2023.126809
9. Chaki J, Dey N, Moraru L, Shi F (2019) Fragmented plant leaf recognition: bag-of-features, fuzzy-color and edge-texture histogram descriptors with multi-layer perceptron. Optik 181:639–650
10. Delval L, Bates J, Jonard F, Javaux M (2025) Field heterogeneity of soil texture controls leaf water potential spatial distribution predicted from UAS-based vegetation indices in non-irrigated vineyards. Biogeosciences 22:513–534. https://doi.org/10.5194/bg-22-513-2025
11. Estrada F, Flexas J, Araus JL, Mora-Poblete F, Gonzalez-Talice J, Castillo D, Matus IA, Méndez-Espinoza AM, Garriga M, Araya-Riquelme C et al (2023) Exploring plant responses to abiotic stress by contrasting spectral signature changes. Front Plant Sci 13. https://doi.org/10.3389/fpls.2022.1026323
12. FAO (2024) The State of Food and Agriculture 2024—value-driven transformation of agrifood systems. Food and Agriculture Organization of the United Nations, Rome. https://doi.org/10.4060/cd2616en
13. Francis LJ, Geissler LGB, Okole N, Gipp B, Stachniss C, Heim R (2025) ReflectDetect: a software tool for AprilTag-guided in-flight radiometric calibration for UAV-mounted 2D snapshot multi-camera imagery. SoftwareX 30:102150. https://doi.org/10.1016/j.softx.2025.102150
14. Frazier AE (2015) Landscape heterogeneity and scale considerations for super-resolution mapping. Int J Remote Sens 36:2395–2408. https://doi.org/10.1080/2150704X.2015.1040130
15. Gabriel A, Gandorfer M (2023) Adoption of digital technologies in agriculture—an inventory in a European small-scale farming region. Precis Agric 24:68–91. https://doi.org/10.1007/s11119-022-09931-1
16. Garg PK (2022) Remote sensing: theory and its applications. New Age International Pvt Ltd., Delhi

17. Garrigues S, Allard D, Baret F (2008) Modeling temporal changes in surface spatial heterogeneity over an agricultural site. Remote Sens Environ 112:588–602. https://doi.org/10.1016/j.rse.2007.05.014
18. González Guzmán M, Cellini F, Fotopoulos V, Balestrini R, Arbona V (2022) New approaches to improve crop tolerance to biotic and abiotic stresses. Physiol Plant 174:e13547. https://doi.org/10.1111/ppl.13547
19. Grünwald NJ, Bock CH, Chang JH, De Souza AA, Ponte EMD, du Toit LJ, Dorrance AE, Dung J, Gent D, Goss EM et al (2024) Open access and reproducibility in plant pathology research: guidelines and best practices. Phytopathology. https://doi.org/10.1094/phyto-12-23-0483-ia
20. Günder M, Yamati FRI, Barreto A, Mahlein A-K, Sifa R, Bauckhage C (2025) SugarViT—multi-objective regression of UAV images with vision transformers and deep label distribution learning demonstrated on disease severity prediction in sugar beet. PLoS One 20:e0318097. https://doi.org/10.1371/journal.pone.0318097
21. Heim RHJ, Melloy P, Ispizua Yamati FR, Okole N, Mahlein A-K (2025) Spatially explicit negative prognosis of Cercospora leaf spot epidemics by process-based integration of leaf area index from remote sensing. agriRxiv. https://doi.org/10.31220/agriRxiv.2025.00351
22. Heim RHJ, Wright IJ, Chang H-C, Carnegie AJ, Pegg GS, Lancaster EK, Falster DS, Oldeland J (2018) Detecting myrtle rust (*Austropuccinia psidii*) on lemon myrtle trees using spectral signatures and machine learning. Plant Pathol 67:1114–1121. https://doi.org/10.1111/ppa.12830
23. Heim RHJ, Carnegie AJ, Zarco-Tejada PJ (2019) Breaking down barriers between remote sensing and plant pathology. Trop Plant Pathol 44:398–400. https://doi.org/10.1007/s40858-019-00300-4
24. Hennecke F, Bömer J, Heim RHJ (2025) Modification of an automated precision farming robot for high temporal resolution measurement of leaf angle dynamics using stereo vision. MethodsX. https://doi.org/10.1016/j.mex.2025.103169
25. Hetzroni A, Miles GE, Engel BA, Hammer PA, Latin RX (1994) Machine vision monitoring of plant health. Adv Space Res 14:203–212. https://doi.org/10.1016/0273-1177(94)90298-4
26. Khanna M, Atallah SS, Kar S, Sharma B, Wu L, Yu C, Chowdhary G, Soman C, Guan K (2022) Digital transformation for a sustainable agriculture in the United States: opportunities and challenges. Agric Econ 53:924–937. https://doi.org/10.1111/agec.12733
27. Kim S, Heo S (2024) An agricultural digital twin for mandarins demonstrates the potential for individualized agriculture. Nat Commun 15:1561. https://doi.org/10.1038/s41467-024-45725-x
28. Kuska MT, Heim RHJ, Geedicke I, Gold KM, Brugger A, Paulus S (2022) Digital plant pathology: a foundation and guide to modern agriculture. J Plant Dis Prot 129:457–468. https://doi.org/10.1007/s41348-022-00600-z
29. Lausch A, Pause M, Doktor D, Preidl S, Schulz K (2013) Monitoring and assessing of landscape heterogeneity at different scales. Environ Monit Assess 185:9419–9434. https://doi.org/10.1007/s10661-013-3262-8
30. Liakos KG, Busato P, Moshou D, Pearson S, Bochtis D (2018) Machine learning in agriculture: a review. Sensors 18:2674. https://doi.org/10.3390/s18082674
31. Mahlein A-K, Arnal Barbedo JG, Chiang K-S, Del Ponte EM, Bock CH (2024) From detection to protection: the role of optical sensors, robots, and artificial intelligence in modern plant disease management. Phytopathology® 114:1733–1741. https://doi.org/10.1094/PHYTO-01-24-0009-PER
32. Mahlein A-K, Kuska MT, Behmann J, Polder G, Walter A (2018) Hyperspectral sensors and imaging technologies in phytopathology: state of the art. Ann Rev Phytopathol 56:535–558. https://doi.org/10.1146/annurev-phyto-080417-050100
33. Mikaberidze A, Cruz CD, Zerihun A, Barreto A, Beck P, Calderón R, Camino C, Campbell RE, Delalieux SKL, Fabre F et al (2025) Opportunities and challenges in combining optical sensing and epidemiological modelling. Phytopathology. https://doi.org/10.1094/PHYTO-11-24-0359-FI

34. Mohanty SP, Hughes DP, Salathé M (2016) Using deep learning for image-based plant disease detection. Front Plant Sci 7:1419. https://doi.org/10.3389/fpls.2016.01419
35. Mridha MF, Dey N (2025) Machine vision in plant leaf disease detection for sustainable agriculture
36. Muszynski M, Klein L, da Silva AF, Atluri AP, Gomes C, Szwarcman D, Singh G, Gu K, Zortea M, Simumba N, Fraccaro P, Singh S, Meliksetian S, Watson C, Kimura D, Srinivasan H (2024) Fine-tuning of geospatial foundation models for aboveground biomass estimation. arXiv. https://doi.org/10.48550/arXiv.2406.19888
37. Natalio AIM, Back MA, Richards A, Jeffery S (2024) Field-scale heterogeneity overrides management impacts following conversion to no-till within an arable system. Appl Soil Ecol 193:105104. https://doi.org/10.1016/j.apsoil.2023.105104
38. Okole N, Yamati FRI, Hossain R, Varrelmann M, Mahlein A-K, Heim RH (2023) Hyperspectral signatures and betalain indicator for beet mosaic virus infection in sugar beet. In: 2023 IEEE international workshop on metrology for agriculture and forestry (MetroAgriFor). IEEE, Pisa, Italy, pp 506–511. https://doi.org/10.1109/MetroAgriFor58484.2023.10424290
39. Outhwaite CL, McCann P, Newbold T (2022) Agriculture and climate change are reshaping insect biodiversity worldwide. Nature 605:97–102. https://doi.org/10.1038/s41586-022-04644-x
40. Pätzold S, Hbirkou C, Dicke D, Gerhards R, Welp G (2020) Linking weed patterns with soil properties: a long-term case study. Precis Agric 21:569–588. https://doi.org/10.1007/s11119-019-09682-6
41. Purcell W, Neubauer T, Mallinger K (2023) Digital twins in agriculture: challenges and opportunities for environmental sustainability. Curr Opin Environ Sustain 61:101252. https://doi.org/10.1016/j.cosust.2022.101252
42. Pylianidis C, Osinga S, Athanasiadis IN (2021) Introducing digital twins to agriculture. Comput Electron Agric 184:105942. https://doi.org/10.1016/j.compag.2020.105942
43. Rajinikanth V, Dey N, Satapathy SC, Kamalanand K (2019) Inspection of crop-weed image database using kapur's entropy and spider monkey optimization. In: Soft computing for problem solving: SocProS 2018, vol 1. Springer Singapore, Singapore, pp 405–414
44. Rodriguesz MA, Cuenca L, Ortiz A (2018) FIWARE open source standard platform in smart farming—a review. In: Camarinha-Matos L, Afsarmanesh H, Rezgui Y (eds) Collaborative networks of cognitive systems. PRO-VE 2018. IFIP advances in information and communication technology, vol 534. Springer, Cham. https://doi.org/10.1007/978-3-319-99127-6_50
45. Roscher R, Russwurm M, Gevaert C, Kampffmeyer M, Dos Santos JA, Vakalopoulou M, Hänsch R, Hansen S, Nogueira K, Prexl J et al (2024) Better, not just more: data-centric machine learning for earth observation. IEEE Geosci Remote Sens Mag 12: 335–355. https://doi.org/10.48550/arXiv.2312.05327
46. Savary S, Willocquet L, Pethybridge SJ, Esker P, McRoberts N, Nelson A (2019) The global burden of pathogens and pests on major food crops. Nat Ecol Evol 3:430–439. https://doi.org/10.1038/s41559-018-0793-y
47. Schramowski P, Stammer W, Teso S, Brugger A, Herbert F, Shao X, Luigs H-G, Mahlein A-K, Kersting K (2020) Making deep neural networks right for the right scientific reasons by interacting with their explanations. Nat Mach Intell 2:476–486 (2020). https://doi.org/10.1038/s42256-020-0212-3
48. Seeley MM, Wiebe BC, Asner GP, Abraham AJ, Cooper HF, Gehring CA, Hultine KR, Allan GJ, Whitham TG, Goulden T et al (2025) Why do models go wrong? The importance of adaptations and acclimations in driving landscape-level spectral variation. Available at SSRN: https://ssrn.com/abstract=5379735 or https://doi.org/10.2139/ssrn.5379735
49. Sharma L (ed) (2020) Towards smart world: homes to cities using Internet of Things, 1st edn. Chapman and Hall/CRC. https://doi.org/10.1201/9781003056751
50. Sharma L (2019) Object detection with background subtraction. LAP LAMBERT Academic Publishing, SIA OmniScriptum Publishing, Latvia, European Union, February 2019. ISBN: 978-613-7-34386-9

51. Sharma L, Carpenter M (eds) (2022) Computer vision and Internet of Things: technologies and applications, 1st edn. Chapman and Hall/CRC. https://doi.org/10.1201/9781003244165
52. Sharma L, Garg PK (eds) (2019) From visual surveillance to Internet of Things: technology and applications, 1st edn. Chapman and Hall/CRC. https://doi.org/10.1201/9780429297922
53. Sharma L, Garg PK (eds) (2021) Artificial intelligence: technologies, applications, and challenges, 1st edn. Chapman and Hall/CRC. https://doi.org/10.1201/9781003140351
54. Sharma L, Garg PK (eds) (2025) Digital twins and simulation technology concepts and applications, 1st edn. Chapman and Hall/CRC. https://doi.org/10.1201/9781003582489
55. Sharma L, Garg PK (eds) (2024) Deep learning in Internet of Things for next generation healthcare, 1st edn. Chapman and Hall/CRC. https://doi.org/10.1201/9781003451846.
56. Sharma L, Garg PK (eds) (2023) Technological prospects and social applications of society 5.0, 1st edn. Chapman and Hall/CRC. https://doi.org/10.1201/9781003324720
57. Sparks AH, Ponte EMD, Alves KS, Foster ZSL, Grünwald NJ (2023) Openness and computational reproducibility in plant pathology: where we stand and a way forward. Phytopathology 113:1159–1170. https://doi.org/10.1094/PHYTO-10-21-0430-PER
58. Tharwat A, Gaber T, Awad YM, Dey N, Hassanien AE (2015) Plants identification using feature fusion technique and bagging classifier. In: The 1st international conference on advanced intelligent system and informatics (AISI2015), November 28–30, 2015, Beni Suef, Egypt. Springer International Publishing, Cham, pp 461–471
59. Toda Y, Okura F (2019) How convolutional neural networks diagnose plant disease. Plant Phenomics 2019. https://doi.org/10.34133/2019/9237136
60. Woodcock CE, Strahler AH (1987) The factor of scale in remote sensing. Remote Sens Environ 21:311–332. https://doi.org/10.1016/0034-4257(87)90015-0
61. Yuan F, Ospina R, Perumal AB, Noguchi N, He Y, Liu Y (2025) Smart agriculture in Asia. Plant Commun 6. https://doi.org/10.1016/j.xplc.2025.101377
62. Zarco-Tejada PJ, Camino C, Beck PSA, Calderon R, Hornero A, Hernández-Clemente R, Kattenborn T, Montes-Borrego M, Susca L, Morelli M et al (2018) Previsual symptoms of Xylella fastidiosa infection revealed in spectral plant-trait alterations. Nature Plants 4:432–439. https://doi.org/10.1038/s41477-018-0189-7
63. Zhang R, Zhu H, Chang Q, Mao Q (2025) A comprehensive review of digital twins technology in agriculture. Agriculture 2025(15):903. https://doi.org/10.3390/agriculture15090903

# Chapter 13
# The Digital Twins Revolution—Are We Ready for It?

**P. K. Garg**

**Abstract** Digital twin is a digital representation of real-world assets or processes that help industries to connect the real world with the digital world. It offers simulation, analysis, and the ability to optimize operation in real-time and reduces operational costs, enhance efficiency and meet sustainability. Digital twins are now transforming manufacturing, healthcare, smart cities, construction, automotive, aerospace, defence, energy industries and other sectors. By constantly gathering and analyzing real-time data from the product or process, the companies can figure out what problems are likely to happen, plan for maintenance, and make things work better. This proactive strategy can help cut costs, boost efficiency, and cut down on downtime. The digital twin revolution has the potential to change how things are done and find new value. This chapter discusses the areas where digital twins have been creating revolutions, challenges in making digital twins dynamic and their future in businesses.

## 1 Background

A digital twin is a digital representation of a physical asset or process that is crucial for changing various industrial sectors. The digital twin technology is functioning as a link between the physical world and digital world for industries to simulate, analyse, and optimize the operations in real-time. Digital twin can be considered a classic approach for businesses who want to minimize operational costs, boost efficiency, and achieve sustainability objectives. Digital twins are now used for more than just demonstrating machines for example, they are used in construction, entertainment, marketing, and customer service, among other fields. In sectors such as manufacturing, healthcare, smart cities, construction, automotive, aerospace, defence and energy, digital twins are rapidly evolving. It assists businesses in maintaining an

P. K. Garg (✉)
Indian Institute of Technology, Roorkee, India
e-mail: pkgiitr@gmail.com



L. Sharma et al. (eds.), *Digital Twin Technology in Geospatial Science: Trends, Applications and Challenges*, Data-Intensive Research,
https://doi.org/10.1007/978-3-032-20311-3_13

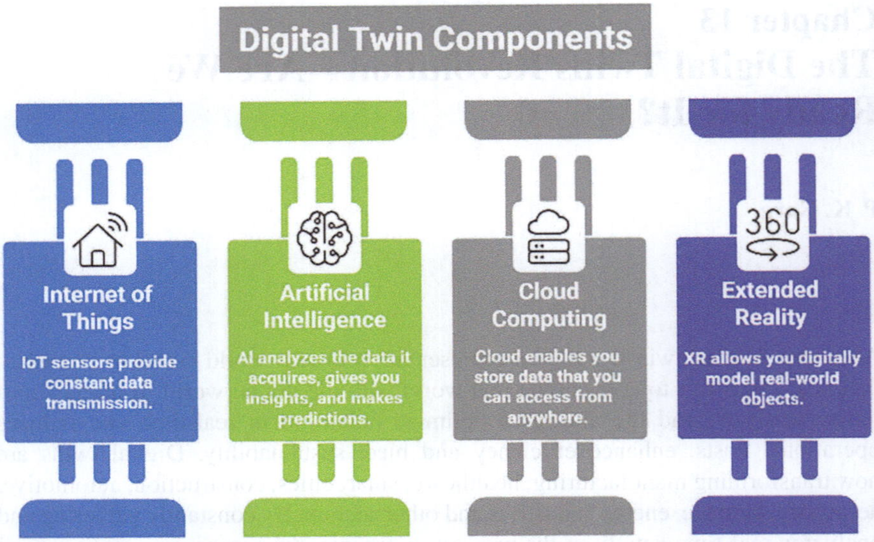


**Fig. 1** Four major technologies supporting digital twins [2]

attractively digital and personal experience 24 × 7. If your firm is based on accuracy and efficiency, digital twins are fostered [15, 16].

In the digital era, companies are able to perform much more effectively, efficiently, and make smarter decisions more rapidly. To do this, they need new age concepts like the most up-to-date and advanced digital solutions like digital twins and Artificial Intelligence (AI). Advanced technologies like AI, Machine Learning (ML), Internet of Things (IoT), cognitive science, Extended Reality (XR), and cloud have transformed the traditional static models of digital twins into dynamic models. The main change with AI technology is that digital twins are now being used in virtual and automated ways. As a result, four main technologies as illustrated in Fig. 1 today drive digital twins together to generate real-time insights, predictive analytics, optimisation, and modern simulations to discovering diversified products of new opportunities and business models.

## 2 Digital Twin Revolution

The initial conceptualization was done in 2002 when Dr. Michael Grieves first introduced digital twins in the product lifecycle management context. The idea focused on creating a virtual model of a real physical asset to optimize processes like design, manufacturing and maintenance. The aerospace and automotive industries quickly adopted it because they saw how it could make things more efficient, especially when it came to simulating complex systems and scheduling maintenance. Although the first digital twin was inducted into the internet of things (IoT) in 2010, the system has come a long way since then Physical assets are embedded with sensors, which transmit real-time data, enabling the digital twin to transition from being static to dynamic and interactive. The establishment of IoT standards and protocols enabled uninterrupted communication between physical and digital assets [1, 13, 18].

The AI and ML were adopted in the 2020s, which enhanced the analytical, predictive, and prescriptive analytics from digital twins. These tools were capable of predicting failures, prescribing solutions, and continuously learning from data. They also made a breakthrough in healthcare, as digital twins of human organs enabled personalized treatment plans and simulations for surgical procedures. Presently, immersive technologies, like Augmented Reality (AR) and Virtual Reality (VR), are making digital twins more accessible and interactive. For example, architects and engineers use AR to visualize building designs in real environments. Immersive interfaces allow for interaction with digital twins intuitively, improving decision-making and collaboration.

In manufacturing, the integration of AI and digital twins provides real-time monitoring, analysis, and optimization. The AI in digital twins can be employed for making predictions for taking near real-time decisions. The AI is helpful in digital twins by analysing Big Data which has been collected from different sources, to assess the maintenance requirements, if any. The processing and analysis of data in real-time using AI enable digital twins to predict future conditions accurately through continuous learning from the processed data [13–16].

The digital twin revolution has already been adopted by many industries, such as:

**BMW**: It is using cloud-based digital twin technology to optimize the production process, and manufacturing system. This enables simulation and optimization by utilizing Azure digital twins. These digital twins have helped BMW improve operations, enhance quality control, and lower down production costs.

**Bosch**: It is implementing cloud-based digital twins in the manufacturing processes by creating digital twins of their production systems. It allows them to carry out real-time monitoring for optimization of operations and enhancement of efficiency.

**General Electric (GE)**: General Electric (GE) is revolutionizing the way industries function by keeping a check on the efficiency of its jet engines, wind turbines and power plants, simulating operational situations, predicting potential issues, and optimizing performance.

**Honeywell**: The company combines real-time data from sensors and machinery to digitally simulate its manufacturing operations via cloud-based technology to enhance their manufacturing processes with digital twin simulations which enables predictive repairs, process optimization, and operational enhancements. This adds a significant value to their customers' value.

**Philips Healthcare**: They create virtual models of patients' organs that allow doctors to simulate surgeries, predict outcomes, and customize treatments according to the patient's unique circumstances to empower patient care.

**Procter & Gamble (P&G)**: Procter & Gamble (P&G) is using cloud-based digital twin technology to improve its manufacturing processes and keep a check on and improve them in real time. The digital twin has made operations more efficient, reduced expenses and made it easier to respond to demands in the market.

**Retail**: It is enhancing the shopping experience for customers by making virtual layouts of stores (like IKEA) and giving customers personalized shopping experiences, both online and in-store.

**Rolls-Royce**: It is implementing cloud-based digital twins to improve its maintenance and servicing operations. The digital twins use data from engines and aircraft systems, and their integration allows predictive maintenance and optimize fuel efficiency. The technology has not only saved the costs but also improved safety and overall performance for its customers.

**Siemens**: Optimize manufacturing using virtual twins to simulate different production scenarios before implementing any real-world changes.

**Tesla**: It is improving the design and performance of its cars by keeping a check on how they work, updating their software, and figuring out when they will need servicing.

**Toyota**: It is working to make driving better by simulating how drivers act and making vehicle safety systems more effective.

**Unilever**: It is streamlining supply chains by optimizing its supply chain and manufacturing processes.

## 3 Digital Twins in Businesses

The digital twins create a virtual replica of a physical object. The digital twins of physical assets are used to monitor real-time data through a network of sensors and simulate behavior in various conditions that result in improved reliability, enhanced sustainability, low costs, and improved productive systems. It helps the industries identify problems from start to breakdown before the actual breakdown occurs. For example, a digital twin of a machine can help predict maintenance needs, which can save money by preventing breakdowns and extending the lifespan of machinery. If the physical system malfunctions, digital twin would help not only in identifying the problems through data and software, but also in reducing closedown time and without much human involvement. In the construction industry, virtual modeling and simulation explore new values of product and infrastructure lifecycles [7].

The digital twins contribute significantly to timely decision-making. Industries can do simulation and predictive analysis by creating a digital twin, for an efficient decision-making. The digital twins have the capabilities to improve decision-making, and products, and optimize maintenance operations. Thus, the virtual, dynamic model of physical assets, or environments that are continuously updated with real-time data, is impacting the "digital twin revolution", allowing improved simulation, monitoring, and optimization, as shown in Table 1.

Efficiency and optimization are the biggest advantages the digital twin offers, as it simulates real-world conditions in a virtual space, allowing for optimization without physically altering the system. The digital twin helps optimize products and production systems, thus saving valuable resources. This is particularly valuable in new product development, as industries can the prototype concepts and designs within a digital environment before manufacturing [10]. The design parameters can be tested and finalised by analyzing real-time data from the physical asset. By simulating new products, processes, or systems virtually, industries can reduce R&D costs and risks associated with actual testing. For example, automotive companies can simulate crash tests on a digital twin before creating the physical prototype. It continuously updates to reflect any change to its physical counterpart throughout the product lifecycle, through feedback in a virtual environment, which helps industries to continuously optimize their products, production, and supply chain processes at reduced cost.

Digital twin is well known for its various applications in the health sector. Similar to early failure predictions and predictive maintenance services, in the health sector, it can be used to predict and early diagnose of diseases that may occur by examining the organs or symptoms in the body. A digital twin of the whole human body can be created to prescribe personalized medicines that bring revolution in the health sector and enhance the improvement rates in the medicated process. Thus, digital twins improve and analyze the predictions for treatment [8].

In health sector, sensors provide data of human body to create digital twins, but this approach can be costly as input information from patients is often obtained by clinical tests, and image scans. But, the data from these sensors improves the diagnosis and treatment processes, enhancing the success rate. Wireless and wearable devices are being designed that will continuously transmit data/information, like heart rate, pulses, blood pressure, etc., to the cloud server. Through software, digital twins of every patient wearing these sensors can be created that would help doctors in emergent situations. Patients can be monitored anywhere, since the smartphone data of patient's health is available on the cloud, and accessible to the doctor. By analysing these data, doctors are able continuously monitor the serious patients, and amend the treatment approach, if required. Wearable sensors allow patients to be tracked physically. Wearable technologies can also be used to examine entire muscular system of patients those who have mobility issues. Medicine therapy, radiotherapy, or surgical operations are only carried out, if digital twins ensures that it is really necessary [5]. The digital twin as a service in Industry 4.0 is proposed in Fig. 2. It lists various smart tools and technologies to create a digital twin, as well as key services that could benefit most in the healthcare sector.

**Table 1** Various functions of a digital twin

| Category | Functions | Key impact |
|---|---|---|
| Quality improvement | Predicts and detects quality defects | Reduces rework and product recalls |
| | Improves overall product quality | Enhances customer trust and product reputation |
| | Ensures traceability to control and resolve quality issues faster | Faster issue resolution and compliance with regulations |
| Operational cost savings | Predictive maintenance avoids unexpected equipment failures | Reduces downtime and maintenance expenses |
| | Optimizes manufacturing processes to enhance equipment efficiency | Improves resource utilization and productivity |
| | Minimizes process variability and operational inefficiencies | Lowers production costs and ensures consistent performance |
| Accelerated product development | Reduces time-to-market for new products by simulating designs virtually | Faster launch cycles and reduced R&D costs |
| | Streamlines the change process with digital models | Better performance and fewer production delays |
| | Tests multiple design iterations digitally before manufacturing | Improves product performance and innovation capabilities |
| Enhanced customer services | Proactively resolves warranty and claims issues using real-time data | Improves customer trust and satisfaction |
| | Efficiently tracks and manages product configurations | Reduces service costs and enhances after sales support |
| | Identifies opportunities for upgrades and tailored maintenance | Higher customer retention and new revenue opportunities |
| Revenue growth | Discovers opportunities for product upgrades and enhanced features | Opens new revenue streams and enhances market share |
| | Offers predictive maintenance and optimization services | Generates additional income through service contracts |
| | Introduces performance-driven business models | Differentiates offerings and increases profitability |
| Lifecycle optimization | Combines "as-designed," "as-manufactured," and "as-maintained" data for a complete asset overview | Better lifecycle management and operational decision-making |

(continued)

**Table 1** (continued)

| Category | Functions | Key impact |
|---|---|---|
| | Extends asset lifecycles by optimizing performance at every stage | Reduces total cost of ownership and enhances efficiency |
| | Meets sustainability goals through resource optimization and waste reduction | Achieves long-term environmental and operational benefits |

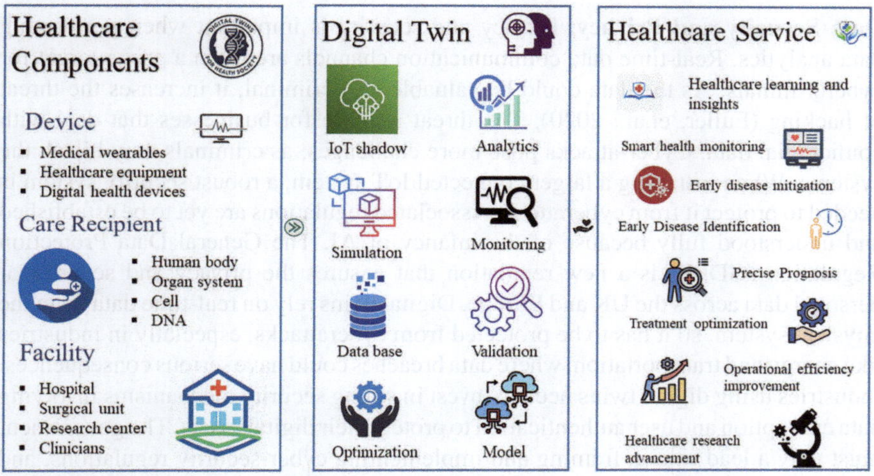


**Fig. 2** Digital twins as a healthcare service [8]

## 4 Challenges Associated with the Development of Digital Twin

Digital twins have great potential to bring revolution in the industries. With their known various advantages, there are several challenges to successful implementation. These data challenges can be addressed to a larger extent with robust data collection and management strategies, and investments in smart infrastructure. The advances in faster digitalization have created the bigger digital divide. In addition, the incorporation of AI technology in digital twins has raised the concerns about equality of access, privacy, ethics, and security [7].

Upgrading infrastructure and enhancing connectivity is crucial as slow internet access limits real-time data exchange. Security and regulatory compliance with rising cyber threats also pose hurdles. Interoperability and standardization is critical for the successful adoption of digital twins. The shortage of skilled professionals in data analytics, and simulation modeling demands more training in the area. Moreover, the initial setup costs can be substantial, especially for small and medium-sized enterprises. Addressing these challenges comprehensively is vital for wide-spread adoption of digital twin technology.

Various challenges associated with health care sector may overcome with the introduction of digital twins. A complete digital twin will be useful to clinical staff, hospital staff, nurses, doctors, and administration. Some of the associated challenges are [2]:

**Data**: The heart and soul of a digital twin is the accurate and updated data. Real-time, and accurate data from various sensors, and simulations are essential to develop an accurate digital twin. It is always a challenging task to get the ideal data ecosystem.

**Data Security and Privacy**: Privacy and security is important when performing data analytics. Real-time data communication channels are often a prime target for cybercriminals. As the data could be valuable to a criminal, it increases the threat of hacking (Fuller, et.al., 2020). The threat is more for businesses that deal with confidential data. Cyber-attacks pose more challenges, as criminals may hijack the systems. When installing a larger connected IoT system, a robust security system is needed to protect it from cyberattack. Associated regulations are yet to be established and understood fully because of the infancy of AI. The General Data Protection Regulation (GDPR) is a new regulation that ensures the privacy and security of personal data across the UK and Europe. Digital twins rely on real-time data from the physical system, so it has to be protected from cyberattacks, especially in industries like energy and transportation, where data breaches could have serious consequences. Industries using digital twins need to invest in strong security mechanisms involving data encryption and user authentication to protect their digital assets. The government must play a lead role in framing and implementing cyber-security regulations, and enhancing security of the system through encryption.

**Data Fragmentation**: Information/data may be available in isolation in various organisations, systems, and formats, which may complicate its integration and analysis. Seamless integration is essential for exploiting their full potential.

**Data Quality**: A geospatial digital twin receives data from multiple sources in different formats, sensors, cameras, LiDAR scanners, etc. The quality of data would generate meaningful business insights. Often, the available data has issues with its quality, such as accuracy and missing values. The quality data should be noise-free. Using inadequate or poor, or inconsistent data in the digital twins, it may lead to inaccurate simulations and predictions as well as underperformance of the digital twin (Fuller, et.al., 2020). This challenge can be overcome by implementing advanced data integration tools that process different types of data and combine them into cohesive output. The data needs to be sorted and cleaned to ensure high-quality data is fed into the AI algorithms.

**Multidisciplinary Approach**: The multidisciplinary approach used for designing and developing digital twins poses other challenges. These challenges occur due to involvement of various specialisations in collaborating. A multidisciplinary approach can help in making breakthrough analysis, but can slow down the progress as several experts think in different directions (Fuller et.al., 2020).

**Integration Complexity**: Making a digital twin completely functional requires a substantial investment in IoT infrastructure, data analytics, software and hardware, cloud services, and AI, including trained manpower. Many companies still face challenges in integrating these technologies seamlessly to create an effective digital twin. Integrating digital twins with limited connectivity and data-sharing capabilities can be cumbersome and costly.

**Real-Time Data**: The main challenge with real-time continuous data is to control the flow of data, and its effective utilisation. The IoT devices will produce large volumes of unstructured data, so sorting and organisation of data is necessary to ensure more data is usable and valuable (Fuller, et.al., 2020). With Big Data, this challenge becomes even bigger. Real-time data capture and transmission may be complex and costly if large assets are to be monitored. Limited bandwidth will also restrict the creation of an effective digital twin. Efficient communication devices, such as 5G, may provide the faster solution. It will connect several sensors and devices, provide high-speed connectivity with greater reliability, and consume ultra-low power, which is needed for faster operation of digital twins. Due to the expenses involved, their accessibility is limited in developing countries. However, the data from a large number of sensors poses complexity regarding data connectivity and processing.

**Connectivity**: Inspite of significant growth in IoT sector, main challenge lies in connectivity, and it is particularly useful when real-time monitoring is done. Various sensors within one manufacturing process may pose a challenge when connecting them together. If one of the sensors is not fully operational, it could affect the data collection process. With a well-connected IT infrastructure, the digital twin will effectively achieve its goals [7].

**IT Infrastructure**: Digital twin needs updated and smart infrastructure for the success of IoT devices and data analytics, as well as facilitates the efficient operation of a digital twin. The biggest challenge is the availability of IT infrastructure, and the cost involved in running these systems. A geospatial digital twin is a complex system of both hardware and software using the latest technology solutions. It requires high initial investments, which may be expensive for small and midsize businesses. In addition, the implementation and maintenance of a digital twin requires hiring specialized experts, which may add to the cost.

The rapid growth of AI requires high-performance infrastructure to handle the huge data and run the algorithms. This challenge may be countered with the use of GPUs "as an on demand service" available through the cloud. For instance, Amazon, Google, Microsoft, and NVIDIA, are among those already provide such services [7]. Use of cloud platform in digital twins throws another challenge in terms of strong security.

**Interoperability**: The lack of standardisation is however a significant challenge to digital twins. Due to lack of any standardization, there is lack of accuracy in the working of digital twins. Efforts need to be made by different stakeholders to develop industry-wide standards that encourage interoperability, quality, and compliance. Therefore, their adoption is limited, and few implementations have been developed to go beyond those that use them, particularly in the manufacturing industry.

**Scalability**: The increase in the amount of data to be processed, it would be challenging to scale the technology to ensure the same level of collection and processing quality. Scaling a digital twin up or out typically involves significant additional resources and may affect operational consistency and component integration.

**Skills gap**: Based on cutting-edge technologies, digital twins need a steep learning curve. Effective implementation and utilization of digital twins needs a specialized team for Big Data analysis, AI, and modeling skills. Training programmes and specialised courses are required to fill this skills gap. The users, looking at the advantages, understand that the AI will instantly save time and money, hence their expectations become high. Thus, the challenges and limitations are to be understood when applying data analytics (Fuller, et.al., 2020). Literacy of AI is essential, allowing potential users how the AI can be applied effectively.

The digital twin technology deploying network services is still in its early stage, and raises several open challenges that are to be addressed. Figure 3 lists the research challenges and future directions for employing digital twins [7].

These challenges necessitate open-source solutions, standard protocols, and finances in upgrading the skills. Further advancements in data management and interoperability will minimize the hurdles of creating and using digital twins, bringing efficiency and innovations in industries.

## 5 Deeper Dive–Future of Digital Twins

The development of digital twin technology has incorporated many objects, like buildings, organisations, and infrastructure. Its adoption in manufacturing, construction, and health sector provides the advantages of low costs and correct decision-making. The digital twin system can collect data about the dynamic assets using IoT sensors [17–20]. Companies can make effective use a digital twin to test a product before it is finalised for manufacturing. Industries can simulate the system to understand unexpected faults and develop appropriate strategies. The digital twins enhance risk analysis, speed up the development of new products, and enhance the reliability of production [2].

While digital twins have been frequently used in manufacturing and aerospace sciences, there is immense potential in future where digital twins will make a dent in healthcare, transportation, retail, energy, etc. From monitoring patients' health to optimising supply chains, the digital twins will enhance innovations in many sectors, providing optimum solutions to complex environments [8]. There is a rise of

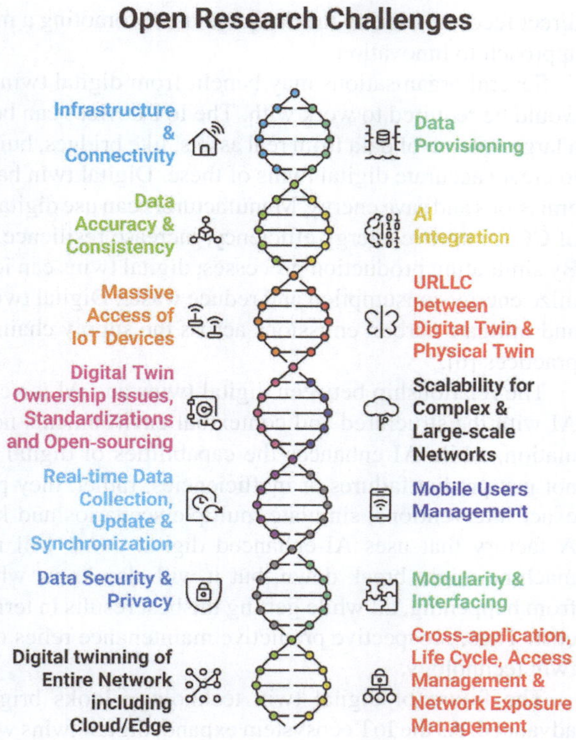


**Fig. 3** Research challenges and future direction for deploying digital twins [7]

digital twins "As A Service" and cloud-based solutions, which will increase accessibility and affordability for various organizations, as it will not require extensive and costly infrastructure, leading to enhanced business opportunities. Cloud platforms will further ensure secure data storage, faster processing capabilities, and easy accessibility [9].

This fusion of AR and VR will revolutionise visualisation, collaboration, and decision-making. It will help in navigating and real-time guidance by integrating digital twins onto physical assets. In addition, it will offer avenues for training, troubleshooting, and remote collaboration. The involvement of edge computing will provide real-time analytics and decision-making, which is important for several applications, like driverless vehicles, smart cities, robotics and automation.

The digital twin is making its mark in the R&D sector, leading to a shift in the way organizations carry out decision-making. Digital twins encourage innovations, as they can bring novelty in economic and social spheres. As an innovation process, digital twins enable collaboration throughout the value chain, facilitating the integration and sharing of data, as well as product development, manufacturing, operation and maintenance. The digital twins also empower users to participate in the innovation process through cloud computing and enable collaboration within the company and with external parties. Further, digital twins allow for real-time monitoring and direct feedback during the design phase, promoting a more efficient and cooperative approach to innovation.

Several organisations may benefit from digital twins, but a large amount of data would be required to work with. The IoT devices can be used to collect and transmit a large volume of data from real assets, like bridges, buildings and production plants, to create accurate digital twins of these. Digital twin has great potential to minimise emissions and save energy. Manufacturers can use digital twins to reduce the emission of $CO_2$, improve energy efficiency, increase resilience, and speed up the processes. By simulating production processes, digital twins can identify opportunities to minimize energy consumption and reduce waste. Digital twins enable companies to track and mitigate carbon emissions across the supply chain. thus promoting sustainable practices [6].

The relationship between digital twins and AI is very strong. Digital twins offer AI with the structured and contextual environments needed to transform the information, while AI enhances the capabilities of digital twins. AI-powered twins do not just predict failures or inefficiencies; rather, they provide recommendations for exact interventions, simulate multiple scenarios and keep learning post-outcomes. A factory that uses AI-enhanced digital twins will not only understand when a machine might break down, but it will also know when, why, and how to stop it from happening, all while getting the best results in terms of energy use, quality, and cost. Thus, prospective predictive maintenance relies on the fast adoption of digital twin technology.

The future of digital twin technology looks bright as it keeps getting more advanced. As the IoT ecosystem expand, digital twins will become more intelligent in the future, leveraging AL and machine learning algorithms. These AI-powered twins can lead to decision-making, becoming more dynamic and self-optimizing systems. Most of the current digital twins depend on a cloud connection for data processing. As edge computing gets better, though, digital twins will be able to process data in real time at the source. This will lead to faster insights and quicker decisions. The integration of IAL and digital twins will offer a platform on which virtual models and real-world applications work together perfectly [9].

The next generation of digital twins exploits advanced AI algorithms and machine learning models to acquire information, adapt, and even automate decision-making without human intervention. This revolutionizes automation in a wide range of fields, ranging from traditional manufacturing to healthcare and smart cities. As a result of this shift, firms are encouraged to innovate and craft strategy rather than be caught up in routine decision-making processes. These twins can respond to real-time changes, such as adjusting supply chain logistics or optimizing energy grids based on demand patterns. With self-learning capabilities, these twins can forecast and prompt potential failures or inefficiencies with greater accuracy [6]. The advantages of next-generation digital twins include design optimisation, predictive maintenance, performance monitoring, and situational awareness. The next-generation networks for digital twins offer improved performance, reliability, scalability, flexibility, and security.

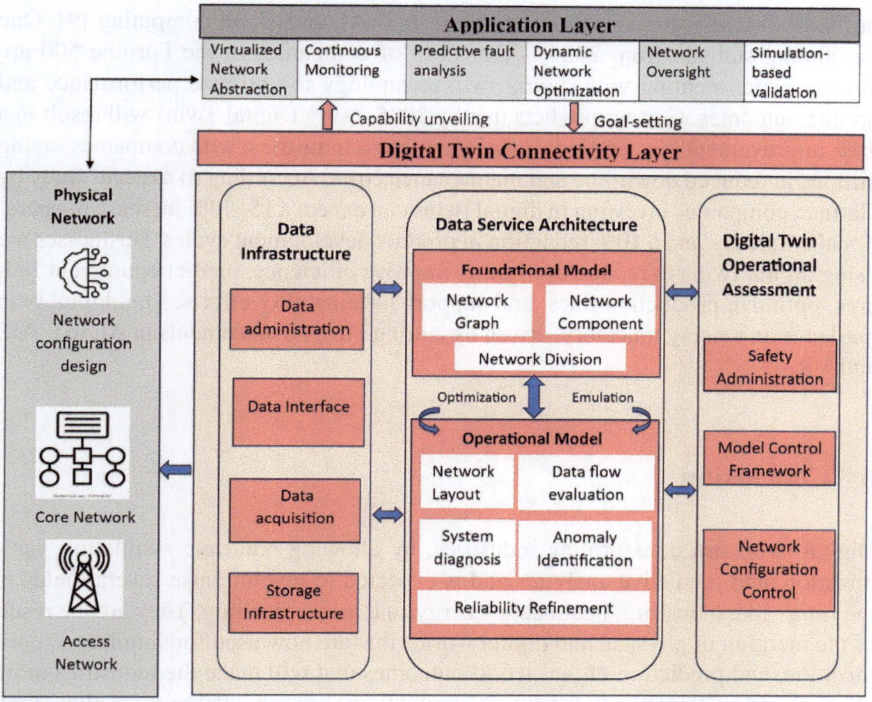


**Fig. 4** Digital twin framework in a three-tier architecture

Figure 4 shows the digital twin framework in a three-tier architecture wherein a closed loop is formed with the physical entities and their counterparts.

The AI revolution in digital twins has made significant progress in the health sector. Patient-specific digital twins can be used to integrate medical history, real-time biometrics, and environmental factors so as to prescribe personalized medicines. Hospitals and administrators may use digital twin models to prescribe a workflow plan for doctors and supporting staff. It helps track appointments, maintain proper inventory processes, and reduce the waiting time of patients. The digital twins of the internal organ systems can be used to suggest medical practice and patient care. Doctors may use this data for organ modelling, testing medicines, and enhancing surgical planning. Digital twins by using AI and IoT sensors offer higher accuracy, consistency, and a complete record of illness and its treatment [5].

The next-generation digital twin will transport data, information, and executable models from development to operation. This information can lead to cost-effective development of systems during operation. The next generation digital twins will inherently have real-time capabilities because they will be built on an event-driven framework to become useful.

The global digital twin market is projected an increase from US $17.73 billion in 2024 to US $259.32 billion by 2032, at a compound annual growth rate (CAGR)

of 39.8%, mainly due to advancements in IoT, AI, and cloud computing [9]. One fascinating statistic from 2024 is that 66% of companies in the Fortune 500 are already experimenting with digital twin technology to optimize performance and predict outcomes. Gartner predicts that by 2025, using Digital Twins will result in a 10% improvement in operational efficiency across industries, with companies saving billions in reduced downtime and maintenance costs. According to a recent study by Gartner, companies investing in digital twins can expect a 15–30% increase in operational efficiency and a 10% reduction in product development cycles. Businesses are using digital twins to reduce downtime, improve efficiency, predict equipment failures, optimize production lines, and support sustainability efforts. The digital twin market is on a rising trajectory, driven by cutting-edge advancements in AI, IoT, AR, and VR.

## 6 Conclusion

Digital twins are transforming industries, by allowing real-time simulation, optimization, and predictive analytics and is expected to revolutionise several fields in the future like industries, healthcare, energy, and urban planning. They are the result of the merging of physical and digital worlds that are now used for simulation, optimization, and prediction of real-world outcomes that will make the industries more efficient and sustainable. In the future, digital twins are expected to grow along with AI and IoT and are set to open new doors for innovation and the way we see or interact with the surroundings. Its adoption will also be on the rise, as more sectors are to benefit from it in the future, and more technologies like healthcare, the retail sector, and construction are to incubate digital twins. Digital twins can be combined with the latest technology like AI, and IoT to maximise their potential for optimization and innovation. They can integrate with several technologies like AR for combining more areas for a better experience and AI techniques and analytics for more effective connections, and analytics [10].

The digital twins can save on operating expenses, improve performance and production tasks, reduce product generation costs, and predict breakdowns. As more and more businesses recognize the potential of the digital twins, a transformation is taking place. As digital twins and related technologies have gotten better, a number of challenges have risen up. While the challenges, which include data security, implementation costs, compatibility with existing systems, are there, the benefits of digital twins ensure that digital twins are an essential tool for any corporation wishing to remain competitive and up-to-date with rapidly changing. As adoption grows, digital twins will play a vital role in determining how industrial and economic development unfolds in the coming years. ‘

Future research incorporating digital twins will concentrate on healthcare domains with substantial potential. Already, the digital twins technology in combination with AI has proven its importance in different areas. The digital twin entirely revolutionizes the old way of monitoring and checking equipment performances, such as using complex analytical tools and ML. As the trend grows and accelerates in the next years, entities will look for new ways to improve their processes, excel in real-time data-based decisions, and invent unique products and services, as well as business design. Digital twins are not just a mirror; they are doors to a world where there is no limit what we can do. The future is knocking—are we ready to embrace the power of digital twins?

## References

1. Ali Z, Biglari R, Denil J, Mertens J, Poursoltan M, Traoré MK (2024) From modeling and simulation to digital twin: evolution or revolution? Simulation 100(7):751–769. https://doi.org/10.1177/00375497241234680
2. Attaran M, Celik BG (2023) Digital twin: benefits, use cases, challenges, and opportunities. Decis Anal J 6. https://doi.org/10.1016/j.dajour.2023.100165
3. Dey N, Wagh S, Mahalle PN, Pathan MS (eds) (2019) Applied machine learning for smart data analysis. CRC Press
4. Dey N, Bhatt C, Ashour AS (2018) Big data for remote sensing: visualization, analysis and interpretation. Springer, Cham, p 104
5. Erol T, Mendi A, Dogan D (2020) The digital twin revolution in healthcare, 1–7. https://doi.org/10.1109/ISMSIT50672.2020.9255249
6. E-SPIN (2025) Digital twin revolution: transform industries with virtual prototyping and innovation in 2025. https://www.e-spincorp.com/digital-twin-revolution/
7. Fuller A, Fan Z, Day C, Barlow C (2020) Digital twin: enabling technologies, challenges and open research. IEEE Access, PP 1–1. https://doi.org/10.1109/ACCESS.2020.2998358
8. Hassani H, Huang X, MacFeely S (2022) Impactful digital twin in the healthcare revolution. Big Data Cogn. Comput. 6:83. https://doi.org/10.3390/bdcc6030083
9. Hendriks R (2024) Whitepaper on the evolution and future of digital twin technology. ICT Group, The Netherlands
10. KaarTech (2024) The digital twin revolution—bridging physical and virtual realities. https://www.kaartech.com/wp-content/uploads/2024/12/The-Digital-Twin-Revolution-Bridging-Physical-and-Virtual-Realitites.pdf
11. Sharma L (ed) (2020) Towards smart world: homes to cities using Internet of Things, 1st edn. Chapman and Hall/CRC. https://doi.org/10.1201/9781003056751
12. Sharma L (2019) Object detection with background subtraction. LAP LAMBERT Academic Publishing, SIA OmniScriptum Publishing, Latvia, European Union, February 2019. ISBN: 978-613-7-34386-9
13. Sharma L, Carpenter M (eds) (2022) Computer vision and internet of things: technologies and applications, 1st edn. Chapman and Hall/CRC. https://doi.org/10.1201/9781003244165
14. Sharma L, Garg PK (eds) (2019) From visual surveillance to Internet of Things: technology and applications, 1st edn. Chapman and Hall/CRC. https://doi.org/10.1201/9780429297922
15. Sharma L, Garg PK (eds) (2025) Digital twins and simulation technology concepts and applications, 1st edn. Chapman and Hall/CRC. https://doi.org/10.1201/9781003582489
16. Sharma L, Garg PK (eds) (2024) Deep learning in Internet of Things for next generation healthcare, 1st edn. Chapman and Hall/CRC.https://doi.org/10.1201/9781003451846
17. Sharma L, Garg PK (eds) (2021) Artificial Intelligence: technologies, applications, and challenges, 1st edn. Chapman and Hall/CRC. https://doi.org/10.1201/9781003140351
18. Sharma L, Garg PK (eds) (2023) Technological prospects and social applications of society 5.0, 1st edn. Chapman and Hall/CRC. https://doi.org/10.1201/9781003324720

19. Solanki VK, Katiyar S, BhashkarSemwal V, Dewan P, Venkatasen M, Dey N (2016) Advanced automated module for smart and secure city. Procedia Comput Sci 78:367–374
20. Tonmoy MR, Hossain MM, Dey N, Mridha MF (2025) MobilePlantViT: a mobile-friendly hybrid ViT for generalized plant disease image classification. arXiv preprint arXiv:2503.16628

MIX
Papier aus verantwortungsvollen Quellen
Paper from responsible sources
FSC® C105338

If you have any concerns about our products,
you can contact us on
**ProductSafety@springernature.com**

In case Publisher is established outside the EU,
the EU authorized representative is:
**Springer Nature Customer Service Center GmbH**
**Europaplatz 3, 69115 Heidelberg, Germany**

Printed by Libri Plureos GmbH
in Hamburg, Germany